The Urban Book Series

AF769088

Series Advisory Editors

Aims and Scope

The Urban Book Series is a resource for urban studies and geography research worldwide. It provides a unique and innovative resource for the latest developments in the field, nurturing a comprehensive and encompassing publication venue for urban studies, urban geography, planning and regional development.

The series publishes peer-reviewed volumes related to urbanization, sustainability, urban environments, sustainable urbanism, governance, globalization, urban and sustainable development, spatial and area studies, urban management, urban infrastructure, urban dynamics, green cities and urban landscapes. It also invites research which documents urbanization processes and urban dynamics on a national, regional and local level, welcoming case studies, as well as comparative and applied research.

The series will appeal to urbanists, geographers, planners, engineers, architects, policy makers, and to all of those interested in a wide-ranging overview of contemporary urban studies and innovations in the field. It accepts monographs, edited volumes and textbooks.

More information about this series at http://www.springer.com/series/14773

Roberto Pasini

Landscape Paradigms and Post-urban Spaces

A Journey Through the Regions of Landscape

Springer

Roberto Pasini
Departamento de Arquitectura
Universidad de Monterrey
San Pedro Garza García, Nuevo León
Mexico

ISSN 2365-757X ISSN 2365-7588 (electronic)
The Urban Book Series
ISBN 978-3-030-08564-3 ISBN 978-3-319-77887-7 (eBook)
https://doi.org/10.1007/978-3-319-77887-7

Softcover re-print of the Hardcover 1st edition 2019

Printed on acid-free paper

This Springer imprint is published by the registered company Springer International Publishing AG part of Springer Nature
The registered company address is: Gewerbestrasse 11, 6330 Cham, Switzerland

Nature: - Thinkest thou then that the world was made for thee? It is time thou knewest that in my designs, operations, and decrees, I never gave a thought to the happiness or unhappiness of man. If I cause you to suffer, I am unaware of the fact; nor do I perceive that I can in any way give you pleasure. What I do is in no sense done for your enjoyment or benefit, as you seem to think. Finally, if I by chance exterminated your species, I should not know it.

—Giacomo Leopardi
Dialogue between Nature and an Icelander
tr. C. Edwardes, Trübner and Co. 1882, p. 78
[1827]

I wrote this book in Mexico.
It is dedicated to all the Adelitas
betrayed by the revolution.

Foreword

The book, *Landscape Paradigms and Post-urban Spaces*, immediately puts its proverbial finger on the existential character and necessity of contemporary urban circumstances bound up with landscapes, namely the inevitable binary that appears to exist, often in multilayered form, between natural circumstances and the artifice of constructed landscapes. To be sure for the author they are multilayered, but it is in the process of unpacking them that some paradigms succeed whereas others fall short. In his opening exploration of natural and man-made systems in contemporary landscapes, Pasini explodes this intriguing multilayering into an intellectually challenging journey that is at times breathtaking, idiosyncratic as it should be and in the end acerbic and insightful. His commentary is never flagging, mundane, or superficial. Moving into the treatment of the composite that emerges, various terms are evoked and concepts are raised, largely from other disciplinary fields for dealing with the inherent two or multisided character of real or imagined overlapping urban landscapes.

Moreover, successive concepts become more complex and intertwined. To begin with, miscegenation derived from the Latin *miscere* and *genus* literally means to mix and was first coined in the United States to refer to inbreeding among people of different races. Back in those earlier days, it usually carried some weight of disapproval of the practice, even if in other cultural settings, this admixing was seen to be potentially strengthening. This is followed by the creole, largely from linguistics and commonly referring to someone from the Caribbean of mixed European and African descent speaking and identified by a language that was stable even if deriving from a mixture of languages—a mother tongue, if you will, formed from the contact of two languages largely through an early pidgin stage, according to most dictionary definitions. Nevertheless, the bifold character of creole does not remain pidgin but evolves into something more complex and the equivalent of a normal language, whereas it transcends a simple hybrid by becoming sufficiently disconnected from a source language. Later on, we come to Foucault's heterotopia and places and spaces that function in non-hegemonic conditions, largely as spaces and places of some 'other'. Certainly, these are places with more layers of meaning in relation to other spaces than meets the mind's eye and are in their dual or

Fig. 1 View of a trail crossing the matorral submontano in the Sierra Madre Oriental of Mexico (*photo* rp)

multiple meanings segues into a necessarily symbolic realm. Towards the end, a test is made of such a symbolic field, including its derivation, through an intriguing construct in the Sierra Madre Oriental of Mexico (Fig. 1). There an itinerary is charted through the natural terrain, with points along it at which travelers are confronted with objects which challenge them by re-framing their perceptions. Consequently, they come to know a place differently and, in a cumulative manner, the trail they are on takes on a multiplicity of novel meanings. The symbolic landscape acts on the natural landscape and *vice versa*.

From beginning to end, this journey satisfies one's intellectual curiosity about its subject, challenges often preconceived ideas and concepts about various forms of landscape 'isms', and finally pushes discussion where it should be in the guise of a project and a symbolic field. As a wise colleague of mine once said when talking about architecture, now I want to talk about the unspeakable or undiscussable.

New York, USA
2018

Peter G. Rowe
Raymond Garbe Professor of Architecture
and Urban Design and Harvard University
Distinguished Service Professor, Harvard GSD

Preface

Brief Synopsis and Structure of the Text

This book explores the merging of natural and man-made systems in contemporary space. It proposes to interpret the entirety of the contemporary continuum through the lens of a landscape paradigm combining scientific and cultural layers. The theoretical elaboration results in the formulation of a 'symbiotic landscape construct'.

The synthetic Introduction of Chap. 1 summarizes theoretical themes and practical issues approached throughout the book and lays out the symbiotic field notion. The book is further organized in three parts. Part I offers a recent panorama of the space-making disciplines dealing with the contemporary continuum. Part II discusses the relevance of the notions of composite and miscegenation in the analysis of the process of spatial formation at a local as well as geographic scale. Part III illustrates a landscape installation implemented in the Sierra Madre Oriental of Mexico as a spatial transcription of the theoretical landscape construct and casts scenarios for future research and design work on the built/natural continuum. Abstracts, keywords, and partial conclusions are provided for each chapter.

Part I

The first part traces a disciplinary panorama of the recent transition from a Greco-Roman conception of spatial organization, based on the idea of confined urban space, to a contemporary condition characterized by the diffusion of anthropic networks over geographic extents. The overview interplays with a bundle of individual and collective trajectories, the ideas of people and groups who have variously engaged the challenges posited by the interpretation of incrementally extensive dwellings on Earth.

The book sets out to expose the relics of obsolete spatial categories and notions in the global discourse on the geographic prospects of the human habitat to sift out the attributes of a novel kind of diffuse 'urbanity' emerging in an all-embracing mixed landscape.

Reconducting theoretical principles to practical terms, a compared anatomy of design cases represents an array of approaches to space-making in the age of the demise of essential divarications between natural and artificial: the cases span from the visionary plans for macro-ecologies being implemented in Asia to the engrafting of micro-ecologies in European cities, the merging of living and artificial structures, and the testing of a possible cybernetic control over natural ecosystems through digital technologies.

Part II

While the urban space suggests series of binaries deriving from the original dichotomy city/nature, the geographic expansion favors the commingling of composite or opposite principles. The second part of the book analyzes ways in which the elective or imposed coexistence of diverse agents in a spatial field kindle a process that evolves from juxtaposition to reciprocal permeation and eventually miscegenation. The research interprets the spatial implications of the colonizing action brought forth by conquering over conquered groups in a cultural and physical terrain, originating processes of confined space-making and broader region-forming.

By adopting the semantic construct of the garden to interpret the outer landscape, the book pursues the exploration of the cavities in-between the tangible topography of the contemporary built/natural continuum and the lacunose territorial mythology that is unawares and fragmentally being produced underneath it. The idea of reformulating the principles for the embedding of a new mythology in the contemporary continuum, its semantization and thereby the reinstitution of a fading sense over the dwelling platform, orients future scenarios.

The research emphasizes the liquid multiplicity of contemporary landscape, referred to as a symbiotic field made of biological and artificial, material and intangible, metabolic and cultural components. Various paradigms used across time to interpret the loose notion of landscape are first dismantled and then reassembled into a symbiotic landscape construct to guide the interpretation and manipulation of contemporary spatial assemblages.

Part III

The third part narrates the physical and spatial transcription of the symbiotic landscape construct into a landscape installation. The 'Tests for a Symbiotic Matorral' project is a set of installations in steel, stone, and wood, forming a 500-m-long floro-faunistic route through the luxuriant *matorral submontano*[1] of the Sierra Madre Oriental of Mexico. Aiming at destabilizing the conventional anthropocentric perspective over the environment, the visitor plunges into a multifocal exploration, entering the perceptive spheres of the many inhabitants of the ecosystem. The installation is also an in situ plotting of the correspondence between

[1]The *matorral submontano* is a low, dense ecosystem extensively covering the hillsides and lower mountain sides of Northern Mexico. Its flora is composed of thicket and scrubland, where a varied fauna finds its habitat. The *matorral* presents an extremely high biodiversity.

Deleuze Guattari's geophilosophical paradigm of 'becoming' and Almo Farina's paradigm of the 'cognitive landscape'. The project, in fact, tests the possible reconciliation of 'cognitive-metabolist' and 'aesthetic-territorialist' landscape models.

The conclusions draw some reflections about the ethic aims and implications of this book and propose an Agenda for the Next Landscape, as a future research and work plan for the space-making disciplines in the contemporary continuum.

Notations for Sources and Quotations

Throughout this book, all sources are credited in detail including authors' and bibliographic data of original texts.

Quotations from other texts are credited and indicated with the following notations:

- "double inverted commas" indicate a direct quotation from the credited text;
- 'simple inverted commas' including a phrase indicate the author's synthetic reformulation of concepts contained in the credited text;
- a citation [e.g., (Rossi 1982, pp. 101–2)] after the final period of a paragraph is referred to the entire period, usually including multiple phrases;
- a citation after a period in the middle of a paragraph is referred to all the preceding phrases of the paragraph;
- a citation before a period is only referred to the specific phrase or part of phrase preceding it.

'Inverted commas' including one or two words are meant to stress a certain expression or notion.

Monterrey, Mexico — Roberto Pasini

Acknowledgements

I would like to thank all the friends who have crossed my path during the years and in particular those who have helped me complete this book in various ways.

This book combines the results of multiple research projects developed during the last 5 years. The preliminary analysis derives from the 2014–2016 TeleTalks conference series curated at the Universidad de Monterrey, Mexico, with some events in collaboration with Laura Cipriani of IUAV Venezia, Italy. I would like to thank Mason White, Lola Sheppard, Kongjian Yu, José Luis Vallejo, Perry Kulper, Alessandro Scandurra, François Roche, Marco Brizzi, Erle Ellis, Ferdinand Ludwig, and Pierre Belanger for their generous availability. The core of the theoretical research presented in this book was elaborated in the frame of the IDAUP, International Doctorate in Architecture and Urban Planning of the Università di Ferrara, Italy and Polis University, Albania, in the period 2014–2017. Special thanks go to my tutors Besnik Aliaj, Rector of Polis, and Gabriele Lelli of DA, Ferrara. I would also like to thank Theo Zaffagnini, for his precious observations, as well as Ledian Bregasi and Loris Rossi for their consideration. 'Tests for a Symbiotic Matorral' is a research project conducted with Patricio Garza, Andrea Ramos, Fernanda Rosas, et al. (see credits in the text) and implemented with the help of, among many others, Rodrigo Legorreta, who manufactured and installed several components, and the distinguished sculptor Jorge Elizondo, who permitted the use of his stone carving ateliers in the Huasteca canyon. The project was brought forth on DIECI research funds of UDEM and in collaboration with the Parque Ecológico Chipinque. I would like to thank the director of research of UDEM Jorge Lozoya for believing in our idea without hesitation. Finally, I am particularly grateful to Peter Rowe of Harvard University and Elisa Cattaneo of the Politecnico di Milano for taking the time to contribute to this book by writing front and back matter.

Contents

Chapter 1
Introduction: Defining a Symbiotic Field

1.1 The Ladder of Rome

'Symbiotic field' refers to the space, *topos* and *chora* (Montaner 2000, p. 99) or *locus* (Rossi 1982, p. 103) and *spatium* (Böhme 2006, p. 403), where anthropogenic and geogenic[1] (Baccini and Brunner 2012) systems intersect, producing the multilayered construct of landscape. Throughout this essay, we will resort to a series of established concepts proceeding from sources diverse in location and time across the geographic configuration of the multidisciplinary debate on landscape to enumerate components of a possible construct. That construct is meant to be climbed upon and then thrown away like Wittgenstein's ladder, which we may call here the 'ladder of Rome', as we are discussing matters related to urbanity.[2] A reduced assemblage made of asserted elements, supporting the constant reformulation of nested frescos, unfolds through multiple layers: the primordial natural platform, *ur-landschaft* (Sauer 1925); its functioning mechanics interpreted by science and environment (D'Angelo 2010a); the superimposed structural and infrastructural network and territory (Camporesi 2016)[3] produced by tangible and intangible fluxes of energy, matter, people, and societal interchange; but also the overall intersubjective and cultural framing of the ensemble (cultural landscape[4]),

---

[1]This metabolist terminology more simply refers to the 'manmade' and 'natural'.
[2]See the 'A corollary note to the Introduction' at the end of this section.
[3]The term *contrade*, used by Piero Camporesi in his book *Le belle contrade* about the 'birth of Italian landscape', is evocative of the transition from a medieval to a modern *entendement* of a land entity that could be translated as 'shires'. Camporesi asserts that the notion of landscape in the Cinquecento was that of the '*paese*', somewhat correspondent to today's notion of territory. See further elaboration on this central issue in Chaps. 2 and 7 of this book.
[4]We have reconstructed in previous publications, the formation of the conventional notion of 'cultural landscape' until its adoption by the highest international institutions, among others UNESCO. Further articulations of the terminology referring to landscape are being coined at sustained pace, such as the notion of 'cultural routes' introduced by the Council of Europe. See: the *European Landscape Convention* of the Council of Europe, Florence 2000; Roberto Pasini, 'Triclini sul mare e rotte culturali', in: *Graphie* n. 66, 2014; Eleonora Berti, *Itinerari culturali del Consiglio d'Europa tra ricerca di identità e progetto di paesaggio,* Firenze University Press 2012.

R. Pasini, *Landscape Paradigms and Post-urban Spaces*, The Urban Book Series,
https://doi.org/10.1007/978-3-319-77887-7_1

Fig. 1.1 Photochrom print of Ludwig II's Schloss Neuschwanstein on the background of the Bavarian Alps, Germany. The Neuschwanstein castle was built between 1868 and 1884 after drafts by Christian Jank, scenic painter and stage designer at Richard Wagner's Bayreuth Opera. It was used as main set for Luchino Visconti's film Ludwig, third episode of his 'Teutonic trilogy'. It is interesting to compare the Neuschwanstein photochrom with the panoramic views of the Grand Budapest Hotel in Wes Anderson's homonymous film of 2014. As detailed in an article by Mekado Murphy, the movie stills of the Grand Budapest Hotel are realized by digitally mounting the foreground of the building's handmade miniature onto the backdrop of a Caspar Friedrich-inspired painted landscape. According to production designer Adam Stockhausen, the authors' aim is "a filtered way of looking at the world" (see: Murphy, Mekado. 'You Can Look, but You Can't Check In', New York Times, 28 Feb 2014). The Neuschwanstein photochrom is a real view selected and processed to coincide with an idyllic landscape fantasized by the Romantic monarch. The Grand Budapest stills are a fictional product assembling fragments of collective imaginary layered and dispersed following the Romantic era. The movie stills aim at providing a displacing experience by immersing the audience into an illusory landscape. In both cases, the landscape results from the combination of real and mental components. (*image* color photo lithograph, author unknown, 1890–1905, PD legal notice at end of chapter)

its interpretation, reflection, and overall vision, that is, the cultural self-awareness of man as a collective inhabiting subject (Fig. 1.1).

1.2 Mixed Space

We live in increasingly mixed spaces. On the one hand, the digital revolution interweaves our reality with a pervasive plot of circuits that produces a material/virtual universe, expanding our consciousness to a new sense of intangible proximity. On the other hand, structural and infrastructural networks extend across open spaces, generating a natural/artificial continuum. We face the need to design planetary macro-ecologies on a geographic scale in order to envision a possible future for our species, while we are urged to reintroduce micro-ecologies in the city space at the human scale to re-naturalize the scenario of our days.[5] 'Cognitive-metabolist' and 'aesthetic-territorialist' approaches to the problem, respectively, preferring environmental engineering versus cultural components of the landscaping construct, have been clashing over the symbiotic field from opposite fronts. In the end, the aggregate result is a general drift towards novel regulations of space based on soft and living systems, casting scenarios for the hybridization or, more radically, the demise of hardscapes.[6]

1.3 Post-urban City

The main purpose of this essay is not to venture out in search of practical planning and urban design strategies to enhance the sustainability of the present order of things. This research rather arises from elemental questions such as: Can a novel landscape paradigm prefigure a space where natural and man-made systems fuse into a synthetic entity? Can it overcome the persistence of explicit and implicit urban-oriented models embedded in the debate on human settlement? With reference to the respective etymological roots of 'city' (*civitas* or the community of citizens) and 'urban' (*urvus* or the area contained inside the curved foundational trench), how can we envision a post-urban city, i.e., a community of citizens, endowed with a novel set of privileges of physical and intangible exchange, dwelling in a non-contained natural/anthropic terrain?

1.4 Centrality of Landscape and the Symbiotic Field

In a time when the natural/man-made dichotomy appears destabilized, the aim of this book is to fundamentally question the irrevocability of the urban and its attributes. The book contributes to the formation of a vision incorporating ground, flora, fauna, and man in a 'symbiotic field', a space where social forms of human dwelling emancipate from anthropocentric perspectives. The vision is supported by

---

[5]Examples are, respectively, the National Ecological Security Pattern, recently adopted by the Chinese government to secure survival of the national territorial system undergoing an unprecedented anthropic pressure and the Air Trees, cyborg-trees designed to reintroduce environmental quality in the metropolitan periphery of Madrid.

[6]In 'Journey Through the Picturesque (a Notebook)', I-ñaki Ábalos and Juan Herreros present "hybrid models [generated by] the interaction between natural and artificial materials" as the basis for a 'new naturalism' that replaces public space with a "hybrid, crossbred, entropic, humanized conglomerate" (Ábalos and Herreros 2003, pp. 56–57).

a multilayered spatial construct instrumentally built up by merging diverse paradigms of the strategic concept of 'landscape'.

An early researcher of the merging of city and countryside into one uninterrupted space (Tunnard and Pushkarev 1963), Christopher Tunnard comes to a precursory understanding of the strategic centrality of the notion of landscape in the interpretation and management of that contemporary continuum: "we must pay special attention to all parts of it as 'landscape' and not isolate the term by thinking of it as applicable only to wild or rural terrain" (Tunnard 1978, p. 116). However, Tunnard's idea of landscape remains firmly anthropocentric, since his adoption of a poetics of 'scenic values' implies a prominence of the human cultural perspective over cognitive diversity. The natural/artificial continuum is, in fact, for Tunnard a 'human habitat'.

The recent widespread concurrence of the design disciplines onto the sphere landscape, and the consequent generation of an extended interdisciplinary debate, offers a propitious momentum to propel our disciplinary repositioning and transitioning from the idea of a smart retrofitting of the obsolete shells of the urban past towards the aim of a symbiotic redetermination of the relationship between man and living space.[7]

The symbiotic landscape construct proposed in this book supports, like the 'ladder of Rome', a surreptitious ascent and an incursion into the more general 'symbiotic field' of the contemporary continuum. The ladder can then be thrown away. What remains is an *entendement* of the mixed surroundings of our days and a scenario for a possible symbiotic future, foreshadowed in the Symbiotic Matorral[8] installation in the Sierra Madre Oriental of Mexico illustrated in the final part of this book.

1.5 A Corollary Note to the Introduction

Wittgenstein says:

> 6.54 My propositions are elucidatory in this way: he who understands me finally recognizes them as senseless, when he has climbed out through them, on them, over them. (He must so to speak throw away the ladder, after he has climbed up on it.) (Wittgenstein 2010, p. 90)

In his philosophical novel *The Name of The Rose*, Umberto Eco rephrases Wittgenstein via his medieval detective William of Baskerville. In an exemplary postmodern literary work, Eco's plot alludes to the rhizomatic nature of contemporary knowledge, where truth does not exist if not as a fragmented bundle of

[7]The symbiotic necessity is a by-consequence of human species' self-proclaimed emancipation from the animal state evoked by calling the product of its own actions artificial, that is, nonnatural.

[8]The *matorral submontano* is a low, dense ecosystem of great diversity. See relative note in Preface and Chap. 8 of this book.

possible routes corresponding to its multiple and unascertained semiotic interpretations. William says:

> I have never doubted the truth of signs, Adso; they are the only things man has with which to orient himself in the world. What I did not understand was the relation among signs. [...] The order that our mind imagines is like a net, or like a ladder, built to attain something. But afterward you must throw the ladder away, because you discover that, even if it was useful, it was meaningless. [...] The only truths that are useful are instruments to be thrown away. (Eco 1984, pp. 287–8)

The same theme of a chaotic reality only explorable through a superstructural hypertext is epitomized in the verse affixed by Adso, the fictional narrator, in his youth William's apprentice, to seal his manuscript: "*stat rosa pristina nomine, nomina nuda tenemus*" ('the rose of old stands in name; we hold mere names', according to the translation by Roland Pepin). As reconstructed by Pepin in his article about the closing line of the novel (Pepin 1986, p. 151), Eco's formulation is an erroneous variant of a verse from Bernard de Morlaix's *De Contemptu Mundi* replacing '*rosa*' for the original '*Roma*', notoriously cited before by Johan Huizinga in his *The Waning of the Middle Ages* of 1924. "*Stat Roma pristina nomine, nomina nuda tenemus*" ('the Rome of old stands in name; we hold mere names') is then the original version, whereby the 'ladder of Rome' is the equivalent of 'Wittgenstein's ladder'.

This book traces routes across the relational carpet that interfaces the most delicate works of nature with the grand cities edified by man—so that the name of a simple rose is interchangeable with that of immortal cities. By virtue of names put on things of the natural and man-made realms, the volatile confessions of man's soul can confront even awful, immutable orographies. Names are, in fact, vehicles for a scientific analysis that does not come at odds with the reveries of an afternoon.

References

Ábalos I, Herreros J (2003) Journey through the picturesque (a notebook). In: Mostafavi M, Najle C (eds) Landscape urbanism: a manual for the machinic landscape. Architectural Association, London

Böhme G (2006) Atmosphere as the subject matter of architecture. In: Ursprung P (ed) Natural histories. Lars Müller, Zurich

Baccini P, Brunner P (2012) Metabolism of the anthroposphere. The MIT Press [1991]

Berti E (2012) Itinerari culturali del Consiglio d'Europa tra ricerca di identità e progetto di paesaggio. Firenze University Press, Firenze

Camporesi P (2016) Le belle contrade. Nascita del paesaggio italiano. Il Saggiatore, Milano [1992]

D'Angelo P (2010a) Estetica Ambientale. Enciclopedia Treccani. Istituto dell'Enciclopedia Italiana, Roma

Eco U (1984) The name of the rose (trans: Weaver W). Warner Books [1980]

Montaner JM (2000) Espacio. In: de Solá-Morales I (ed) Introducción a la Arquitectura. Ediciones UPC, Barcelona

Pasini R (2014) Triclini sul mare e rotte culturali. Graphie (Il Vicolo) (66)

Pepin R (1986) Adso's closing line in the name of the rose. Am Notes Queries 24:9–10

Rossi A (1982) The architecture of the city. The MIT Press, Cambridge [1966]
Sauer CO (1925) The morphology of landscape. Univ Calif Publ Geogr 2:19–53
Tunnard C (1978) A world with a view. Yale University Press, An Inquiry into the Nature of Scenic Values. New Haven
Tunnard Ch, Pushkarev B (1963) Man-made america: chaos or control?. Yale University Press, New Haven
Wittgenstein L (2010) Tractatus logico-philosophicus. Introduction by Bertrand Russell (trans Ogden CK). Project Gutenberg [1921]

Legal Notice:

Fig. 1.1: Unknown (https://commons.wikimedia.org/wiki/File:Neuschwanstein_Castle_LOC_print.jpg), „Neuschwanstein Castle LOC print", marked as public domain, more details on Wikimedia Commons: https://commons.wikimedia.org/wiki/Template:PD-1923

Part I
A Disciplinary Panorama

Chapter 2
Metropolitanism, Its Filiations, and Its Consequences

Abstract By analyzing a series of territorial paradigms proposed by theorists, critics, and designers, this chapter offers a critical overview of the recent transition from metropolitanist to regionalist perspectives over today's expanded dwelling space in the global debate. The analysis tackles metropolitanist models related to Manhattanism and neo-nomadism with reference to their *avant-garde* and radical parentages. The text critically investigates models derivative of metropolitanism, such as landscape urbanism, urban age, and post-metropolis, along with models that present themselves as alternative, such as sub-urbanism and mega-regionalism. The text confronts metropolitanist models and their filiations with the principles of the disciplinary refoundation based on the idea of urban space, such as those of Tendenza, and with the traditional constructs of territory and landscape, such as those of the Italian landscape and the trio of natures, which they propose to subvert. The theoretical elaborations of landscape urbanism are then confronted with their geophilosophical references. The text proceeds to observe the process of expansion of the idea of city over the geographic scale. A novel form of expanded urbanity, surfacing from the analysis, finally suggests the opportunity to identify a set of novel attributes relative to the citizenship of the contemporary built/natural continuum. The conclusions identify the idea of a territorial mythology, unawares and fragmentarily building up in the contemporary continuum, as a possible field of future research and design work.

Keywords Metropolitanism · Manhattanism · Sub-urbanism · Landscape urbanism · Picturesque · Mega region · Paris · French revolution Urban age · Middle landscape

2.1 Introduction

The first 15 years of the new millennium have seen a laborious repositioning of the general interest in the field of design, switching from metropolitan glamor (Fig. 2.1) to mega-regional environments. Richard Florida's controversial reading

R. Pasini, *Landscape Paradigms and Post-urban Spaces*, The Urban Book Series,
https://doi.org/10.1007/978-3-319-77887-7_2

of *The Rise of the Mega Region* (Florida et al. 2007), completing his theory of the 'creative class' (Florida 2002), intercepted this shift in the disposition of the wider architectural audience.

2.2 A Critique of Metropolitanism

2.2.1 *Metropolitan Spell*

The spell of the 'metropolitan promise' had begun to propagate in 1978, riding on the planetary success of a retroactive manifesto for Manhattan, *Delirious New York*, and relying on Rem Koolhaas' rhetoric ability, lyric puissance, and analytical potence. The aloof announcement of the "imminent segregation of mankind" into the tribe of 'Metropolitanites', graduates of the gigantic locker-room skyscraper, multitasking 'metropolitan bachelors' self-redesigned at the price of sterility, and "the remainder of human race" (Koolhaas 1978, pp. 152–60), had seduced the undifferentiated multitude.

For the Metropolitanites, Koolhaas manufactures a new groundbreaking typology, the Manhattanist Skyscraper, by means of a critical reinvention operated on the Downtown Athletic Club. If Koolhaas is the deus ex machina establishing Manhattan's missing mythology, his Skyscraper is the *machina* set on the theater of critique for the congestions of metropolitanism to burgeon and the individuals to surrender to "the definitive instability of life in the Metropolis" (Koolhaas 1978, p. 157). However, Koolhaas' lyric enacts the individual's surrender to compulsory dissatisfaction and obsession for material and immaterial consumption, alongside Manhattan's 'floor by floor conquest of the Skyscraper by social activity'.

In a rhetoric apotheosis, the American way of life, significantly made of 'know-how'[1] and 'initiative' (i.e., capitalist entrepreneurship), "definitively overtake[s] the theoretical lifestyle modifications that the various 20th-century European *avant-gardes* have been insistently proposing, without ever managing to impose them." Koolhaas' typology, the Manhattanist Skyscraper, is explicitly identified with a "Constructivist Social Condenser [that is] a machine to generate and intensify desirable forms of human intercourse" (Koolhaas 1978, p.152). Using Benjamin de Cassares' words, Koolhaas celebrates "the black mass of Materialism" of the Metropolitanites, 'male to the core', who 'divinize matter, energy, motion, change'.

The wishful future of a captivated global design scene is, by that, cast into a dense, congested, and dazzling space: the 'metropolitan' estate.

[1]Cfr. Ernst Jünger's reflection on the notion of 'technique'.

Fig. 2.1 Looking east from Shin-Marunouchi's mezzanine: Tokyo's 1914 Central Railway Station on the glazed backdrop of the business district (*photo* rp)

2.2.2 *Machina diaboli*

Referring to Lévi-Strauss, John Berger comments on how the widespread success of oil painting from the sixteenth century on responded to the identification of 'seeing' with 'possessing'. Buying an oil painting allows the owner to possess the things it represents, such as "landscapes, women, food, dignitaries, mythology" (Berger 1972, pp. 83–4), which results unlikely with both ancient frescoes and modern murals. According to Berger, in ancient oil painting as in modern media "the 'ideal' spectator is always assumed to be male and the image of the woman is designed to flatter him" (Berger 1972, p. 64). If "a man's presence is dependent upon the promise of power which he embodies" and exerts on external subjects, by contrast "a woman's presence is intrinsic to her person", recalling in a very delicate analogy "an almost physical emanation, a kind of heat or smell or aura" (Berger 1972, p. 46). We could conclude that Koolhaas' Skyscraper is a machinic system of possession exerted upon the matter and the woman, animate and inanimate subjects, a *machina diaboli* that perpetrates domination.

2.2.3 *Lenin's byt*

Molded as an anti-utopian palimpsest erected on Constructivist grounds, Koolhaas' Skyscraper for the metropolitan bachelor is presented as a frontal attack on the *petit-bourgeois* social structure based on the family unit. In *An Archeology of Socialism*, Victor Buchli analyzes Moisei Ginzburg and Ignaty Milinis' Narkomfin communal housing in central Moskow (1928–32) as the epitome of the 'social condenser', that is a spatial arrangement inducing the implementation of Vladimir Lenin's agenda of social and feminist emancipation for the achievement of the *byt*, the Soviet lifestyle (Buchli 1999, pp. 63–5).[2]

The achievement of the *byt* by means of implementing standardized communal housing types had been commissioned by the Russian Republic to Ginzburg and other architects associated to the Constructivists groups in 1928. The Narkomfin is the most prominent example among only six such buildings implemented before the

[2]The idea of flatly sweeping out *petit-bourgeois* households through new collective housing types constituted the architectural embodiment of the theories for a collectivist reformulation of life that permeated the *avant-garde* art milieu in the years before and after the Bolshevik revolution (Buchli 1999, pp. 63–76). A more capillary campaign opened another front inside the *petit-bourgeois* household. Coping with the lack of budget to plan the replacement of the actual structures, a series of housewife's manuals were meant to sweep out at least the *petit-bourgeois* values from the domestic sphere, associating hygiene and taste for clean order with socialism (Buchli 1999, pp. 41–2, 52–5). On Alexander Bogdanov's and Alexei Gastev's collectivist theories, see also Anatole Senkevitch Jr.'s 'Introduction' to Moisei Ginzburg's *Style and Epoch* (Senkevitch 1982, pp. 29–31).

Stalinist counter-directives reject the full socialization of the living pattern in favor of the preservation of the *petit-bourgeois* 'domestic heart', that is the familial structuration of society.

2.2.4 Stalin's VOPRA

The ends of the *byt* reform would soon then be called 'leftist' or 'Trotskyist' with the consolidation of Joseph Stalin's power in the Central Committee of the Communist Party during the First Quinquennial Plan (1928–32). The projects are silenced and dissolve along with the Russian constructivists, one of the most talented generations of modern architecture. Vittorio De Feo is probably the first author to propose an articulate study of Russian Constructivism in *URSS Architettura 1917–1936* (Officina 1963), reconstructing the trajectory of Soviet *avant-garde* architects starting in 1920 from the Muscovite art ateliers of the VKhUTEMAS and the competing MVTU.[3] It is no coincidence that VKhUTEMAS' Higher Art and Technical Studios are established by Lenin's direct decree to pursue the same emancipatory agenda in the crafts and industrial production and management. The progressive programs of Nikolai Ladovsky's ASNOVA (Association of New Architects) and Ginzburg's OSA (Organization of Contemporary Architects) are rooted in those ateliers. By 1932, their 'deterritorializing' action will be already suppressed at the hands of Arkady Mordvinov's VOPRA and then Union of Soviet Architects. The reactionary agenda of the association, created in tune with the tyrant's purpose of normalization, is based on a neo-academic monumentalism as a design parallel to the refoundation of the conventional family structure at the core of societal stability (De Feo 1963, pp. 29–45, 55–8).

2.2.5 Manhattanist Condensers

Koolhaas' seminal manifesto of Metropolitanism is structured on the assertion/denial model, consisting in first asserting and then denying daring critical ideas via flamboyant lyric figures. After having celebrated the implemented apotheosis of a Constructivist socio-spatial revolution through the reformulation of the

[3]Among the major figures active in the field of architecture at the VKhUTEMAS, worth of mention are Vladimir Tatlin, Nikolai Ladovsky and his 1928 graduate Georgy Krutikov, El Lissitzky, Kostantin Melnikov, Ilya Golosov, Berthold Lubetkin, Alexander Vesnin, Leonid Vesnin, Moisei Ginzburg and his 1927 graduate Ivan Leonidov, while engaged at the MVTU Alexander Kuznetsov and Victor Vesnin, although many were active in both institutions. For details about the structure of academic programs at VkHUTEMAS, see: monoskop.org/Vkhutemas.

Manhattanist social condenser, the illustrated account of 'The Story of the Pool' at the closing of *Delirious New York* (Koolhaas 1978, pp. 307–10) represents diametrically opposite conditions of daily life in Manhattan, at odds with the ingenuous Constructivist ideals. Koolhaas depicts the imaginary journey of a community of Constructivist architects seeking escape from the Stalinist regime in the early '30s, onboard a visionary floating pool capable of auto-motion designed in the radical revolutionary milieu a decade before. For the special locomotion system of the pool, the Constructivists swim backwards towards America, reaching Manhattan 40 years later to find out that the same crude uniformity they are fleeing from has soaked Manhattan's capitalist society. So, they take off again in an endless anarchist's drift. Koolhaas' apologue of the pool, just like his chant of the Manhattanist social condenser, is an attempt at rooting his rhizomes deeper down into the corpse of Constructivism and steal its soul.

The arborescent model of society founded on the base unit of the family, had been attacked by the Constructivists on behalf of Lenin's revolution just to be surreptitiously readopted a couple of years later for the edification of Stalin's monumental eversion. Similarly, Koolhaas' pseudo-constructivist social condenser does not accommodate liberated socialist individuals of both genders, but rather aggressive bachelors, obsessed with the exploitation of woman as a natural service within the severe frame of entrepreneurial depredation.

2.2.6 Of New Nomads

Accounting of the early appearance of American suburbia along the expanding metropolitan railways and motorways in the 1920s, Peter Rowe links the phenomenon to the "radical restructuring of households [...] during the progressive era (1890–1920)", which had comprehensively replaced the extended Victorian family structure with nuclear units of simplified hierarchical complexity in the fabric of the urbanizing American society (Rowe 1991, p. 4). In fact, the diffusion of the city beyond its urban boundaries, is from the beginning, even though inadvertently, linked to a thorough social restructuration of metropolitanist character.

Toyo Ito's Pao 1 and Pao 2 dwelling units for the Tokyo Nomad Woman (Ábalos and Herreros 1995, pp. 32–7, 47)[4] of 1985 and 1989 respectively, elaborate on the same concept of a molecular metropolitan society where 'home' is replaced by 'pod'. If 'home' implies rootedness and place, 'pod' recalls a technical support installed on an incidental location. Ito's *paos* are 'anti-homes' designed for a mutant society. Tokyo woman's nomadism is supported by light platforms serving dried-up

---

[4] In El Croquis dedicated to *Toyo Ito 1986 1995*, see in particular the section 'The Tokyo Nomad Woman'.

functions such as 'style', 'snack', and 'intelligence', depowered projections and occasional episodes derived from the fundamental acts of hygiene, nourishment, and culture. Resting is the only fundamental act to be supported by the *pao*, while any other primary activity is performed in the collective facilities variously dislocated in the metropolitan continuum, daily crossed by the nomadic trajectories of metropolitanites. If Ginzburg's Narkomfin collectivizes basic functions such as cooking, eating, studying, leisure within a community aimed at a reformed society of liberated individuals, Ito's *paos* disintegrate the identitarian activities of the metropolitanite into a diffuse metropolitan space.

"In any age, a dream for a new life leads into a new space" and the *paos* "depicted an image of an urban life which daily loses reality in proportion to the rate of visualization of the city life"—says Ito (Ito 1995, p. 10). In fact, Ito's *paos* narrate the city drifting from reality towards simulation, while sedentary urbanity becomes nomadic and female, and paternal arborescence becomes rhizome (Ito 1995, pp. 6–15). The metropolitan nomad, however, is not re-achieving a pristine state previous to that of sedentary, but rather a machinization of the human body and its acts symmetrically opposite to the machinization of the human body performed by Koolhaas' Down Town Athletic Club. Ito's metropolitan machinization of the body is a vaporous female diffusing in the atmosphere,[5] while Koolhaas' machinization is a voracious male living for the sole cult of matter.

2.2.7 *Of Other Nomads*

In his 1954 movie *La Strada*,[6] Federico Fellini narrates the journeys of a vagabond strongman and the fragile, credulous young woman he has bought from her wretched mother for 10,000 Italian Lire. Zampanó and Gelsomina traverse the countryside and the mountains of the Italian immediate post-war, a preindustrial world littered of haggard villages, peopled with sparse peasants, publicans, and nuns, stray dogs and gaunt hens, immersed in a constant wintry light. Their poor belongings are packed on a derelict Davis tricar, onto which Gelsomina casts the attributes of a conventional home. "I have in it everything like in a house"—says Gelsomina to a nun, taking naïve pride on the domestic items packed in the tricar's caisson.[7] Zampanó, instead, sleeps indifferently indoor or outdoor stretched out amidst the weeds and the debris of the wasteland.

---

[5]With reference to woman's presence defined by Berger as a "heat or smell or aura", see sub-chapter 'Machina diaboli'.

[6]*La Strada* is an Italian movie of 1954 directed by Federico Fellini, screenplay by F. Fellini, T. Pinelli, E. Flaiano.

[7]Scene of the convent form *La Strada*.

Zampanò is a visionary assemblage, made possible by Fellini's filming technique of 'numerological diction'[8] combining the brutish corpulence of Anthony Quinn and the cavernous voice of Arnoldo Foà. Zampanò is a myth that shares some fundamental characteristics with Ernst Jünger's venerable figure of the *Waldgänger* (the one who goes back to the woods),[9] mauled by the hailing of expanding technique, which *Storm of Steel* and *Sturm* describe in form of the war crossing the world. Like the *Waldgänger*, Zampanò is thus cast into a mythological world, regressing back before the dwelling of sedentary civilizations, to recover the pristine state of nomadism. To that state, he forcefully drags naïve Gelsomina, who wastes away in a constant pallor and ends up destroyed.

The mythical nomad Zampanò stands in sharp contrast with the metropolitan nomads. He is antipodal to the Manhattanist bachelor as to the Tokyoite woman. Zampanò is equally incompatible with the radical figurations that precede metropolitanism: No-Stop City's neo-nomads, who walk across a comprehensively technified world with no need for a motive system, as well as Supersurface's liberated hippies that roam around over their endless vivifying grid, wearing beat clothing and growing beards, enjoying merry company and fraternity. The neo-nomads are freed, emancipated from work and any practical obligation, master magic technology and their lifestyle lysergic.[10] Zampanò has no privilege of election: he struggles to fulfill primary needs. He has no privilege of fairness: he steals, cheats, and brutalizes. Zampanò is not entitled to freedom, he drifts along an anarchic route throughout untamed chaos.

2.2.8 Site and Program

In his *Sub-urbanism and the Art of Memory* of 2003 Sébastien Marot describes Koolhaas' approach as 'super-urbanism' entailing a complete role reversal between

---

[8]The 'numerological diction' is a filming technique that mounts in studio the audio tracks of effects and voices onto mute footage. The method allowed Fellini to direct a diverse cast of international actors that would roam the scene spelling out nothing but numbers or speaking each in their mother language. He could even intervene live in the scene shouting orders and suggestions through a loudspeaker. Fellini took large advantage of the virtuosities of the audio technicians at Cinecittà, arousing acrimonious criticism from American colleagues, who considered his method unorthodox.

[9]In an interview to Antonio Gnoli and Franco Volpi, Jünger asserts: "Ho cercato di mettere a fuoco tutto ciò nel saggio *Der Waldgang* (*Trattato del Ribelle*), in cui il Ribelle, l'Anarca, «passando al bosco», cioè ritirandosi nei penetrali di sé stesso, affronta e vince l'angoscia, il dubbio e il dolore. [...] Dal punto di vista dell'Anarca, del grande Solitario, totalitarismo o democrazia di massa non fanno molta differenza. L'Anarca vive negli interstizi della società, la realtà che lo circonda in fondo gli è indifferente" (Jünger 2006, pp. 53–5).

[10]About the 'erotic manipulation' of space in Superstudio as well as in Loos and de Bastide, see: sub-chapters 'Unsolicited attentions', 'Baker is the landscape', and 'Uneven parallels' in Chap. 5 of this book.

the major project categories of 'site and programme' (Marot 2003, p. iii). The 'great question' in the design field posed by Alberti of "how to choose a site where a city or a programme will be built" (Marot 2003, p. i) has been transformed into "the programme [...] envisioned, shaped and built as site" (Marot 2003, p. iii). Faced with the uncontested primacy of super-urbanism, Marot launches the 'sub-urbanism' alternative with a plea for a symmetrical role reversal whereby "the site becomes the regulatory idea of the project" (Marot 2003, p. iv, note 1) without resurrecting *passéiste* formulae.

2.2.9 *Memory and Verticality*

As much as super-urbanism operates on the metropolitan, sub-urbanism operates on the 'third territorial state' between city and country, the suburban. To derive a series of 'heuristic principles' applicable to design work on this third condition, Marot draws on Frances Yates' investigation of the ancient art of memory[11] and John Dixon Hunt's work on the history of the art of gardens.[12] As ancient rhetors used to associate a sequence of concepts to places within an imagined architecture or anthropic landscape, the art of garden can be regarded 'as a medium for the semantization of the land, lending to nature the status of landscape' by means of plotted correspondences of "a fountain to a spring, a grotto to caves, [...] the parquetry of flowerbeds to the divided plots of cultivated fields" (Marot 2003, p. 20). A point by point correspondence of the imprint of thought and action, interpretation and manipulation, analysis and design, characterizes Marot's sub-urbanist approach, whereby the 'heuristic principles' or 'precepts' of memory and depth elaborate on both the temporal and spatial stratification of the site (Marot 2003, p. iii). This weft of projected correspondences eventually precipitates into the idea of the 'verticalization of landscape' in multiple layers, vertically reorganizing entities traditionally arranged along a horizontal distribution (urban, political, agricultural, wild/divine nature, timeline). The verticalization of landscape through superimposed geo-memory layers is the core of a long-gestated sub-urbanist manifesto, *Palimpsestuous Ithaca*, enwoven by Marot over several years as a specular narrative to Koolhaas' retroactive manifesto for Manhattan.[13] For Marot, the geological stratification deep down in the profundity of Ithaca's glacial lake is the specular counterpart to the programmatic congestion building upwards on top of Manhattan's bedrock.

[11]Marot is referring to Frances Yates, *The Art of Memory*, Routledge and Kegan Paul 1966.

[12]Marot is referring to John Dixon Hunt, *L'Art du jardin et son histoire*, Odile Jacob 1996.

[13]See online: Marot's F.L. Olmsted Lecture at Harvard GSD 2010; Berlage Public Lecture Program at TU Delft 2013.

2.2.10 *Horizontal Syntax*

A counterpart to the metropolitanist vision, New Urbanism, sees its charter published in 2000 (Leccese and McCormick 2000). Since 1993, the new urbanists have assembled retrograde visions in analogy with a Christopher Alexander's 'pattern language' (Alexander 1977), arranged according to the three scales of 'region: metropolis, city, and town', 'neighborhood, district, and corridor', and 'block, street, and building'. In attacking the sprawling of American suburbia and infrastructural dispersion, new urbanism calls for the consolidation of a totally 'horizontal' hierarchy, a backward-looking network of urban nodes with syntactical structure (Yaro 2000, p. 22), and a return to past figurations on the ground (Rossi 1982, p. 51). New urbanism's horizontal hierarchy forges re-enactments of the 'persistent plan' mentioned by Rossi with reference to Marcel Pöete's work. The recipe of 'new urbanism' can be regarded as an adapted version of the traditional syntax of urban space, expanded over the horizontal dimension to the point that any vertical articulation results irrelevant. The loss of vertical articulation denotes the disappointing insubstantiality of a counterfeited solution.

Landscape urbanism and new urbanism can be interpreted as antipodal perspectives within the same field of an asserted novel ruralist, socially aware, and ecological sensitivity, which is starting to erode the metropolitanist aura by exploring space from a 'regionalist' perspective.

2.3 A Critique of Landscape Urbanism

2.3.1 *Machinic Landscapes*

The idea of the landscape as a framework, the productive interrelation between multiple layers of interdependent variables, involving first of all the nature/artifice systemic commingling and material interfacing, the interaction of cultural imprints and environmental mechanics, the poietic concepts, the locale and their assembled in situ transcription are, in fact, the novelty of 'landscape urbanism' announced in James Corner's *Recovering the Landscape* of 1999.[14] The idea of landscape urbanism will be later broken down into practically identifiable actions in Moshen Mostafavi and Ciro Najle's manual for the machinic landscape (Mostafavi and Najle 2003, pp. 6–9). There, Mostafavi's introductory essay cursorily summarizes

[14]The work investigates "the inevitable constructedness of landscape (that is neither natural nor given) and how productive reciprocities among ideas, representations, and physical spaces may be better understood" (Corner 1999, p. ix).

its characteristic points, further elaborated in essays by other authors, such as that of the dissolution of the distinction between city and countryside, and those of the temporality of the ever-incomplete landscape opposed to the implicit finitude of zoning, the landscape as a framework of imagination, the territory as a dialogue between buildings and landscapes where each term is simultaneously present to be, or to be construed as, the other. Furthermore, the combination of urbanism and landscape by way of "transposition of techniques and vocabulary [...] on a metaphoric and metonymic register" produces a multilayered aggregate of cultural, social and political agents interacting with the 'formal and aesthetic performance'. Finally, the idea of the landscape's process-based temporal relativeness (the ephemerality of instant configuration inseparable from the long span of the overall succession) and the idea of the operational productiveness adopted from the venerable practice of agriculture (where visual appearance is under-arched by functioning) engender the demise of formal contextualism. The shift is from typological manipulation to management of processes and relational assemblages.

2.3.2 *The Role of External Forces*

However, one of the most debatable issues brought over by Mostafavi as the very core of landscape urbanism's proposal is the necessary 'openness' to "the role of external forces in the shaping of our cities", forces such as "the opening and closing times of international financial markets" (Mostafavi and Najle 2003, p. 9). The impact of the international financial markets in the shaping of cities would show its drawbacks after the explosion of the 2007 crisis generated by the planetary financialization of territorial transformations. The concept of 'openness to external forces' is constructed upon the logics of the infinite growth model applied to the field of space production on the wake of the uncritical embracement of Wall Street's and the City's neo-hedonism of the early 2000s, a diffuse attitude among the design academia and practice pushing to partake in the globalization banquet. The New York High Line urban regeneration initiative, led by the *Friends of the High Line* civic association and propelled by the Diller Scofidio + Renfro and Corner's project, was saluted worldwide as the highest achievement in the virtuous rehabilitation of abandoned urban heritage in metropolitan environments. It now becomes interesting to reassess such an enterprise, combining the virtues of societal activism, public space generation and private profitability, not simply for its generic reconciliation of public and private interests or grass-root and design-driven processes, but also on the basis of a more articulated analysis of the redistribution of its benefits over the material and social fabric at a regional scale beyond the metropolitan ambit. That is reconsidering the operation in terms of the intensification of public amenities in privileged environments parallel to the dispersal of by-consequences over the suburban area.

2.3.3 Athens

Such initiatives, identifiable with "the alternative models of urbanism [...] open to [...] participation by all citizens" demising "the nostalgic yearning for lost models of public space, monuments, piazzas" (Mostafavi and Najle 2003, p. 9) mentioned in Mostafavi's conclusions to the manual, explicitly spell out the diametrical collision of the principles of landscape urbanism with the modes of construction of the architecture of the traditional western city. We should therefore investigate whether this "redefinition of the public sphere" is fit to "set the scene (albeit momentarily) for democracy in action" as announced.

By advocating the shift "from the deliberate manipulation of typological configurations [...] to the systematic management of virtually open relational assemblages", Najle clearly formulates landscape urbanism's proposed alternative, representing a novel 'browsing across contingency' as opposed to a traditional 'ideological positioning', with the intention to bury Rossi's theory of the city.

However, the idea of the High Line as a Manhattanist *agora* of novel modes of democracy falls short of articulation compared to the Arendtian exploration of *logos*-action space of the Greek *polis* (Arendt 1998, pp. 192–8),[15] as well as of auto-poietic depth before Aldo Rossi's epics on the mythical manifestation of public space in Pericles' Athens (Rossi 1982, p. 131).

2.3.4 Elitist Strolls, Autocratic Recreations

In this regard, we could assess the comprehensive virtues of the High Line enterprise (1999–ongoing) with the virtues of the Lion Mountain project by Tom Leader Studio of 2016. On the one hand, the elitist civic activism of Manhattan residents opens the way to bulks of private profits along the non-programmed trajectory of a design of contingencies that yields a hyper-park for Metropolitanites[16] and tourists. On the other hand, a large scale territorial transformation, heavily top-down programmed by the institutions of an authoritarian regime and hetero-directed by a design platform based on the other side of the Pacific Ocean, operates on the obsolete structures of a run-down amusement park in the peripheral area of a Chinese conurbation. The Lion Mountain project includes the recreation of picturesque landscapes combining a mountain and an artificial lake, the creation of a series of open or enclosed artificial biomes controlling both flora and fauna, a number of varied activity zones recovering visual connection to lost geographic

---

[15]See in particular 'The Greek Solution', in section 'V. Action'.

[16]Regarding the earlier elitist tribe of sterile metropolitan bachelors described by Koolhaas as intent on redesigning their own body (see sub-chapter 'Metropolitan spell'), we can interpret the redesign of the High Line sparked by a participative process as a self-redesigning action, applied to the inherent habitat of the group.

features. The project also reconstitutes cultural continuity with the *loci* of an ancient lifestyle enshrined in surviving painting, such as the Prosperous Suzhou scroll. Elaborated by designers with no relationship to the resident community, the project tends to reconnect the alienated mega-city agglomeration with its orographic substrate at the geographic scale. An archeological reconstruction/reinvention of the city's lost cultural heritage is the bearing idea of the project, that is the reinstating of a supporting mythology alternative to the unrooted drives of finance.

2.3.5 Dissensus or Barbarism

Despite the open reference to Félix Guattari's 'relational assemblages', the attempted 'agorafication' of the High Line does not mark any divarication from the phenomenon of globalization. In his *The Three Ecologies*, in fact, Guattari introduces the concept of 'relational assemblages' to propose a possible 'social ecology' in sheer response to globalization (Guattari 2000, p. 60). 'Living autopoietic machines' (Guattari 2000, p. 61) formed by groups of individuals perform "processes of continuous resingularization [that makes them] more united and increasingly different". That is a social phenomenon of 'heterogenesis' (Guattari 2000, p. 69)[17] aiming to enact "a multifaceted movement, deploying agencies [...] and dispositives that will simultaneously analyse and produce subjectivity" (Guattari 2000, p. 68). The 'autopoiesis of subjectivity' is for Guattari destined to install itself in the 'realms of environment', 'social and institutional assemblages', and 'landscapes and phantasies of the most intimate spheres of the individual' (Guattari 2000, p. 69).

The Italian biologist Almo Farina links globalization to a 'loss of diversity' and consequent production of 'heterogeneity' (Farina 2009, pp. 143–53). Gilles Clément describes the same phenomenon as a new Pangea condition incumbent on the planet due to the vertiginously increasing mobility of humans (Clément 2015, p. 26). Yet, the apparent increasing richness within an area, due to unintentional or planned dissemination of alien seeds from every corner of the Earth, cannot compensate for the real decreasing bio-diversity of the planet. The increasing local bio-diversity of any area and the decreasing overall bio-diversity, in fact, are the two sides of the general loss of order of the planetary ecosystem due to globalization, tending towards entropy, chaos, heterogeneity, that is background noise.

[17]It is worth mentioning that by 'heterogenesis', Guattari refers to the 'increment of diversity' counteracting globalization and not to the opposite notion of 'heterogeneity', used in biology to mean 'chaos' and 'entropy' in a background that becomes indistinct. See Chap. 7 in this book.

It is by 'cultivating *dissensus*' (Guattari 2000, p. 50) towards the 'Integrated World Capitalism (IWC)',[18] whose contemporary manifestation is identified with globalization, that the autopoietic subjectivity can rehabilitate the three spheres of environmental, social, and mental ecology.[19] *Dissensus* towards IWC is *dissensus* towards globalization as to a novel barbarism. Therefore, the 'relational assemblages' mentioned by Najle, despite his implied references, don't resemble Guattari's "collective assemblages of enunciation" (Guattari 2000, p. 61) (syntactical assemblages of autopoietic subjectification). Najle's 'relational assemblages' are rather aligned with the opposite agency of the Guattarian system, the "imaginary crowd aggregates" (Guattari 2000, p. 60) of standardized subjectivity generated by IWC's conversion to syntax production.

2.4 A Critique of the Urban Paradigm and Its Territory

2.4.1 Of the Continuity of the Production Matrix in the Territory

The idea of a territory supported by "a [recognizable] structure of invariant formal matrices"[20] capable to produce 'durable wealth' advocated by Alberto Magnaghi and the *scuola territorialista* (Magnaghi 2010, p. 299), is in fact aligned with Guattari's description of the Japanese and Italian virtuous models of 'reindividualization' by means of "grafting high-tech industries onto a collective subjectivity, while retaining ties to a sometimes very distant past". Here, individualized collective subjectivity is inseparably implanted into the persisting territorial 'matrix' of "certain archaic features of the pre-capitalist era [that] have been inherited and maintained" (Guattari 2000, p. 63).

Magnaghi's formulation of "a structure of invariant formal matrices" supporting the territory certainly draws upon Rossi's idea of the permanence of form in urban artifacts. However, the long-term productive quality installed upon that structure evokes the stability of an Arcadian condition in history.

---

[18]According to Guattari, IWC, or post-industrial capitalism, tends to move away from the production of goods and services in favor of the production of signs, syntax, and subjectivity (Guattari 2000, p. 47) standardized in three tiers including serial subjectivity of the salaried, mass subjectivity of the uninsured, and elitist subjectivity of the executive (Guattari 2000, p. 61).

[19]Guattari calls 'ecosophy' the complex of the three ecological spheres (Guattari 2000, p. 41).

[20]Translation by the author from the Italian original.

2.4.2 *Of the Continuity of the Spatial Substance in the Industrial City*

According to Rossi, the vision of a permanent Arcadia falls into the stream of 'romantic socialism' traced back to William Morris and the Modern Movement that naively consider "the problem of modern urbanism [...] as determined by the city's historical relationship to the Industrial Revolution". Where an Arcadian vision evokes the problem of the modern city as the problem of the industrial city, per Rossi, instead, "the problem of the large city" precedes the industrial period and is, in fact, "the problem of the city itself". He maintains, in fact, the principle of the essential continuity of urban problems, that is the refusal of any change of quality between classical, ancient, modern, and post-modern spatial conditions. Friedrich Engels is quoted with respect to this very point as an inerrable authority, while his dialectic materialism is the compass that orients the whole of Rossi's theory of the city (Rossi 1982, pp. 154–5).

2.4.3 *City as Field of Application of Economic and Political Forces*

The historic chain of violent downfalls and ruthless upheavals accumulated onto the city's material register, including demolitions, reconstructions, urban reforms, equal for their destructive effects to war and bombings, are nothing but "accidental occurrences" capable to "accelerate certain tendencies that already exist" and permit "a more rapid realization of intentions that are already present in economic form" such as "building, acquiring, and selling of land". The underlying dialectic of the 'constant forces that unfold throughout history according to occasional directions offered to them' is a dialectic of "normal economic forces". A shift in the quality of the city, that is the appearance of a distinct spatial stuff, might only have a chance to occur in the transition from the capitalistic to a perfect Socialist city induced by the abolition of the private property. But Rossi's analysis is engaged and rooted in the existing state of things, not projected into a possible future (Rossi 1982, pp. 141–4).

The application of economic forces to the structure of the capitalistic city is, in fact, manifested in "speculation", which largely determines the city's growth and ultimately its 'form' in combination with another underlying agency, that of the 'political choice'. The 'political choice' can be identified with the 'plan', which throughout Rossi's reflection is attributed an autonomous self-agency. The political choices in fact are converted into 'plans that impose themselves with the force of an urban artifact', or the "deeds of certain individuals whose wills acted as historical forces". Political forces converted into plans and deeds join the economic forces to shape the underlying dynamic of the urban formation (Rossi 1982, pp. 140–2).

2.4.4 Speculation as Vehicle of Trans-historical Aspirations

Rossi's notion of 'speculation' regarded as the application of economic forces to the process of urban formation, relies on Maurice Halbwachs' analysis of the deep transformations impressed by the transition from *ancien régime* to modernity in the eighteenth and nineteenth centuries.[21] Halbwachs punctually reconstructs the reform of Paris' layout of boulevards, avenues, and streets, implemented across the whole of the nineteenth century, originating from visions that started surfacing as far back as under the reign of Luis XIV. Those visions are then systematically (although incompletely) formalized during the revolutionary regime, partially implemented under the First Empire, continued during the Restauration and the July monarchy, to be monumentally completed under the direction of Baron Haussmann during the Second Empire.

Formations of trans-historical, even more than collective, aspirations for a novel spatial order hover catalyzing until circumstantial conditions occur to offer the chance to precipitate. The large land acquisitions operated by the revolutionary regime after the decrees for the nationalization of clerical and monarchic property and for the confiscation of *émigré* property between 1789 and 1792, with consequent sales and redistributions over the following two decades, are the gate that channels the materialization of the urban reform of Paris and consolidate the practice of expropriation for the action of the Prefect of Seine in the third quarter of the century. According to Halbwachs, the term 'expropriations', beyond its plain legal notion, comes to represent a more general class of land property dynamics driven by economic forces caught in the act of their materialization into urban space. Rossi consequently resorts to 'expropriations' to refer to 'purely economic phenomena' producing 'total acts' of urban transformation, whose specific circumstances are irrelevant (Rossi 1982, p. 143). More generally, we could think of 'expropriation' as the deeper dynamic driving a chronologic trajectory to perturb a spatial surface, by which history precipitates into geography.

2.4.5 The Monographic Studies on the Trans-historical Aspirations of a City

In 'Les plans d'extension et d'aménagement de Paris avant le XIXe siècle Paris' of 1920, Halbwachs analyzes the ambitious Plan des Artistes pour Paris proposed by the temporary commission of artists of the revolution in 1793 to find that it results surprisingly coherent with the urban renovation implemented by Haussmann in the

[21]Rossi refers to *Les expropriations et les prix de terrains à Paris (1860–1900)* published by Cornély in 1909 and its second extended edition *La population et les tracés de voies à Paris depuis un siècle* published by Presses Universitaires de France in 1925.

1850s and 1860s (Halbwachs 1920).[22] What is a spatial manifesto of the seize of power enacted by the bourgeoisie over the *ancien régime* is implemented by a 'regime promising the extension of material prosperity while practicing the restrictions of political rights' (Rossi 1982, p. 142). Haussmann's vast program of boulevards in Paris, in fact, can be considered the formal occurrence exhibiting the economic hegemony of that same bourgeoisie over the working class (Rossi 1982, p. 142), sanctioned by the massacres of the Communards in 1871.

The implementation of the urban axis of Rue de Rivoli, is the epitome of Halbwachs' urban thesis. Halbwachs had already reconstructed how plans for the reform of the urban front north of the garden of the Tuileries can be traced back to Louis XIV (Halbwachs 1909, 1920). The Plan des Artistes of the Revolution, which had just guillotined Louis XVI, revives those intentions and plans an urban axis on the north side of the gardens from Place de la Concorde, alongside the Palais du Louvre. From there, the axis shifts south to realign itself with the front façade of the *palais*. The façade of the *palais* becomes the west terminal of a grand boulevard extending eastwards up to the monumental circus of Place de la Bastille by cutting through the medieval fabric of Paris. The monarchic and revolutionary plans would then be implemented with secondary variations at a later phase, when the institutions that had traced them had been overturned.

Napoleon will in fact extend eastwards the codes imposing a uniform façade that had already been set half century before over the private buildings on the north front of Place de la Concorde. The constructions on the north side of the Tuileries gardens up to the front of the Louvre are then completed according to a similar urban uniformity, celebrating then name of one of the emperor's first victories against the Austrians at Rivoli. The restored monarchies and the Second Empire will complete the urban axis from the Louvre onwards. Haussmann will revise the impracticable alignment proposed by the Plan des Artistes to more simply extend the existing segment overriding the winding alleys to join Rue Saint-Antoine. The new alignment will more practically take advantage of the ecclesiastical expropriations of the revolutionary period. Solid masses will be pierced and reshaped into façades to eventually reconnect the Concorde to the Bastille.

Moreover, Rue de Rivoli, ideally connecting the Bois de Boulogne with the Bois de Vicennes, becomes the central segment of the east–west urban axis of the *grand croisèe*, the cross that refounds the urban structure of a novel Paris according to Haussmann's monumental schemes (Halbwachs 1920, pp. 12–3). Halbwachs concludes that:

[22]Halbwachs' essay, appeared in the journal La vie urbaine, is largely based on the 1913 *Aperçu historique* edited by Louis Bonnier and Marcel Pöete gathering the work of the Commission de l'Extension de Paris of the Prefecture de la Seine, in-charge of the preparatory work for the city plan competition of 1919. The *Aperçu* studies the movements of the population of Paris towards the suburbs in the nineteenth century compared to the same phenomenon in other French cities as well as London and Berlin. Pöete is also founder of the journal La vie urbaine.

> … les tracés de voie et le changements de la estructure superficielle de Paris s'espliquent non point par les desseins concertés d'un ou plusieurs individus, par des volontés particulières, mais par des tendances ou besoins collectifs auxquels les constructeurs, architects, préfets, conseils municipaux, chef d'État ont obéi, sans prendre, de ces forces sociales, une conscience bien claire, et, quelquefois, avec l'illusion qu'ils s'inspiraient de leurs conceptions propres.[23] (Halbwachs 1920, p. 20)

Transitioning from rationalism to modernity via empiricism, from aristocracy to bourgeoisie via revolutions, from monarchy to empire via revolutionary republic, the urban transformations (or expropriations) of the city constitute the material register of French history in general, epitomized by the accumulation of monuments on the east-west axis of the *croisèe*, therefore *axe historique*, from La Bastille to La Villette.

2.4.6 *Dialectic Materialism and Globalization*

The acceptance of the violence of the historic process over the territory characterizes Rossi's vision, where radical degradation opens the way to spatial reformation. The 1859 plan Cerdá for Barcelona exemplifies for Rossi an 'expropriation', in which the economic forces of speculation are offered the opportunity to fulfill their aims only once the plan lends itself to a profound degradation. The original plan, a technological vision 'too advanced for its time' informed to an 'untenable' low density, has to convert its *illes* into massively constructed blocks, 'magnificently lending themselves to the aims of speculation', in order to produce the unparalleled urban environment of Barcelona (Rossi 1982, p. 150). It is a 'total act' of transformation freeing the flow of 'normal economic forces' regardless of the miserable compromises of its 'accidental occurrences'. Rossi's vision of the economic forces unfolding over the urban space is construed on the same dialectic materialism that will inform the endless deterritorialization and reterritorialization cycles, rewriting the geographic canvas of Deleuze Guattari.

Rossi's idea of speculation as the motor of urban formation, on the one hand, reaches beyond accidental circumstance and toward the dynamic of fundamental forces of a non-pulse history in the process of becoming geography, in a leap analogous to Deleueze Guattari's tension towards the Plan(e) of Immanence. On the other hand, Rossi's dialectic materialism renounces to face the expansion of globalization and remains bound to pre-globalization conditions, under which the negotiation with nature is localized and unique even in the sprawling suburbs. The city remains to Rossi a 'fact in nature' and a 'work of art', that is the greatest good,

[23]In English: "… the street layout and the changes in the superficial structure of Paris are explained not at all by the practical designs of one or more individuals, not by personal wills, but rather by collective tendencies or needs, which the builders, architects, prefects, municipal councils, heads of state have obeyed, without apprehending, of those social forces, a clear awareness, and, sometimes, with the illusion to be inspired by their own conceptions." [translation by the author]

the "human achievement *par excellence*".[24] Tellingly, in the introduction to the American edition of *The Architecture of the City*, he claims that "American architecture is above all 'the architecture of the city': primary elements, monuments, parts" (Rossi 1982, p. 15). "Amorphous zones do not exist in the city, or where they do, they are moments of a process of transformation [...] as in the suburbs of the American city", awaiting "redevelopment". His explanation of the American city is that of English matrixes transplanted into the New World, noting by Peter Hall that the "idea of the precinct" has been practiced for centuries in London's colleges and inns of court, up to Patrick Abercrombie's post-war plan for the peripheral detouring of through traffic for the Westminster and Bloomsbury areas (Rossi 1982, p. 95).

2.5 A Second Critique of Landscape Urbanism

2.5.1 A New Ethos for the Machinic Landscape

Despite Mostafavi and Najle's overall edition, the ideology of the 'machinic landscape' is construed in the essays by Christopher Hight and James Corner. Hight elaborates on several fundamental concepts that define the identity of landscape urbanism through transdisciplinary contagion between architecture, planning and landscape design.

Among these concepts, central is the replacement of the ethics of architecture, operating through 'stasis, truth, wholeness, and timelessness', and the ethics of planning, based on 'control, determinism, and hierarchy', with an ethos of landscape design, operating through 'temporality, complexity, and soft-control' (Hight 2003, p. 24).

The reformulation of a new sense of temporality of the project in the direction of the management of an open, dynamic, and impermanent process, rather than the definition of a completed configuration fixed forever, is probably one of the highest contribution of landscape urbanism to the recent discourse on the contemporary

[24]Rossi refers to "human achievement *par excellence*" (Rossi 1982, p. 33); or "human creation *par excellence*" (Rossi 1982, p. 57); or "human thing par excellence" (Rossi 1982, p. 112, 163). Here, Rossi is rephrasing Lewis Mumford defining the city "a fact in nature [and] man's greatest work of art" (Mumford 1938, p. 5) in conjunction with Claude Lévi-Strauss aphoristic formulation "La ville ... la chose humaine par excellence" from p. 122 of the classic 1955 Plon edition of *Tristes Tropiques*. Rossi discusses the two mentioned definitions respectively in notes 1 and 2 to Chap. 1 of his book (Rossi 1982, p. 180). *Tristes Tropiques* assembles the recollections of Lévi-Strauss' anthropological, as well as auto-analytical, explorations in South America (Lévi-Strauss 2012). In the same book, Lévi-Strauss recounts of the aborted foundation of a fortified colony predating Rio de Janeiro by the hand of a riotous French/native bunch led by Nicolas Durand de Villegagnon. The construction of Fort Coligny off the Bahía de Guanabara will end up catastrophically in an extenuating turmoil of somber insanity and reciprocal treacheries. The story forms a grotesque pendant in relation to the triumphal claim of excellence rendered to the construction of cities.

continuum. Succession mapping-based plans such as Michel Desvigne's Thirty-Year Planting Development for Guyancourt (1989), centered on the natural succession of the ecosystem, opened the way for Corner's elaborations on temporality and certainly constituted the foundations for projects by phases such as Field Operation's Downsview Park in Toronto (1999) and Fresh Kills Park in New York (2003–ongoing).

2.5.2 *Simulacrum-Depiction and Model-Encoding*

Another significant replacement of traits proper to the ethics of architecture with those of landscape that is claimed by Hight to landscape urbanism relies on the project's contamination with the scenographic character of landscape as opposed to the idea of the project as a 'model of order'. The reference in this case is to the inferiority attributed by Reyner Banham to the landscape designer *vis-à-vis* the architect, consequent to the ethically subordinate condition of the act of 'depicting an order' to the act of 'operating on the order' (Hight 2003, p. 25).

As a result of the ethical contamination, the design of the phenomenological simulacrum can be admitted into the field of action of the novel architect dealing with softness and impermanence across time. The depiction of the world is after all an act of analysis, a way of organizing the magma of reality into an order. To this respect, Corner's role in emancipating the landscape architecture practice from a conventional twofold-indirect agency is evident.

The origins of landscape design are conventionally described as following a twofold-indirect agency, orchestrating a scenery by imitating a *mimesis* of a natural context as depicted in a painting. As a multiverse agent, the emancipated landscape architect, conversely, controls the eidetic-strategic hybridization of 'simulacrum depiction' and 'model encoding' (Hight 2003, p. 27), eventually acting on both ends of the landscape construct: what in Farina's cognitive model is described as the observer and the ecosystem (i.e.,the 'cognitive' and 'physical' properties of landscape) (Farina 2009, pp. 15–6). The opposition 'depiction of the simulacrum' versus 'encoding of the model' again refers to the Deleuze-Guattarian duality of the series tracing-arborescence-history versus mapping-rhizome-geography (Deluze and Guattari 1987, p. 296) and that of the 'plan(e) of transcendence' (subjectivities and identifications) (Deluze and Guattari 1987, pp. 265–6) versus the 'plan(e) of immanence' (intensities and speed) (Deluze and Guattari 1987, pp. 254–5), whose plotted interface casts a comprehensive reflection of the world.

We will come back to the construct of the two plan(e)s further on, as it results fundamental in the merging of scientific-cognitive with geophilosophical-territorial paradigms. As of now, we can identify a proportionality between the series of 'tracing/subjectivity/horizontal-organization-of-facialities/simulacrum depiction' and 'mapping/intensity/horizontal-continuum-with-degrees/model-encoding' (Deluze and Guattari 1987, p. 296), leading from a picturesque to a metabolist approach to the contemporary continuum promoted by landscape urbanism.

The most relevant contribution from landscape urbanism to the contemporary discourse on landscape is certainly the consolidation of this idea of temporality and dynamic formation both through theoretic and critic elaboration.

2.5.3 *Of Planes and Rotations*

The same coherence cannot be found instead in Hight's discourse advocating for the rotation of the operational plane of design from a vertical to a horizontal alignment. On that rotation, the 'orientalization' of architecture through landscape as its 'adjacent other' (Hight 2003, p. 23), attributed to landscape urbanism, finds its foundation. As we propose to illustrate, what Hight presents in the core of his essay as a reformulation of the subjectivity of architecture operated by landscape by way of rotating the project's organizational plane from the asserted vertical alignment of architecture to the horizontal alignment of landscape, appears to be the exact opposite. That operation represents, in fact, the adoption of the traditional organizational horizontality of architecture into the practice of landscape. A colonization of landscape of such a sort operated by architecture, corresponds to the imposition of a conquering regime (reterritorialization) while excluding a specular destabilization of the conqueror's condition (deterritorialization) deriving from the exposition to the conquered.

Hight starts from Walter Benjamin's intuition of the 'longitudinal and transversal cuts through the world's substance' (Fig. 2.2), intended as, respectively, the pictorial (picturesque) representation that encloses things and the symbolic graphics (diagramming, mapping) that encloses signs (Hight 2003, p. 29). Canonic art theory, says Higth referring to Rosalind Krauss, has generally identified the longitudinal cut with the verticality of painting's canvas and the transversal cut with the horizontality of graphic printing's flatbed (Fig. 2.3).[25] A geometric discrepancy is yet generated when Krauss uses Benjamin's powerful intuition to backup her equally powerful image of the transition from canvas to flatbed, as per Benjamin, both longitudinal and transversal cuts are conspicuously operated by vertical planes, one longitudinal and the other transversal (Fig. 2.4). The correspondence between the two visions can be one of proportion, while their actual geometries diverge. On that geometric discrepancy is actually elaborated Hight's notion of the 'rotation of planes'.

2.5.4 *Of Planes and Masochism*

Hight, in fact, follows to draw a parallel with Deleuze Guattari's description of the landscape as made of a horizontal expanse (*milieu*) of vertical facialities stating that

[25]*McLuhan in Space: A Cultural Geography*, Richard Cavell proposes the same consideration (Cavell 2002, p. 125).

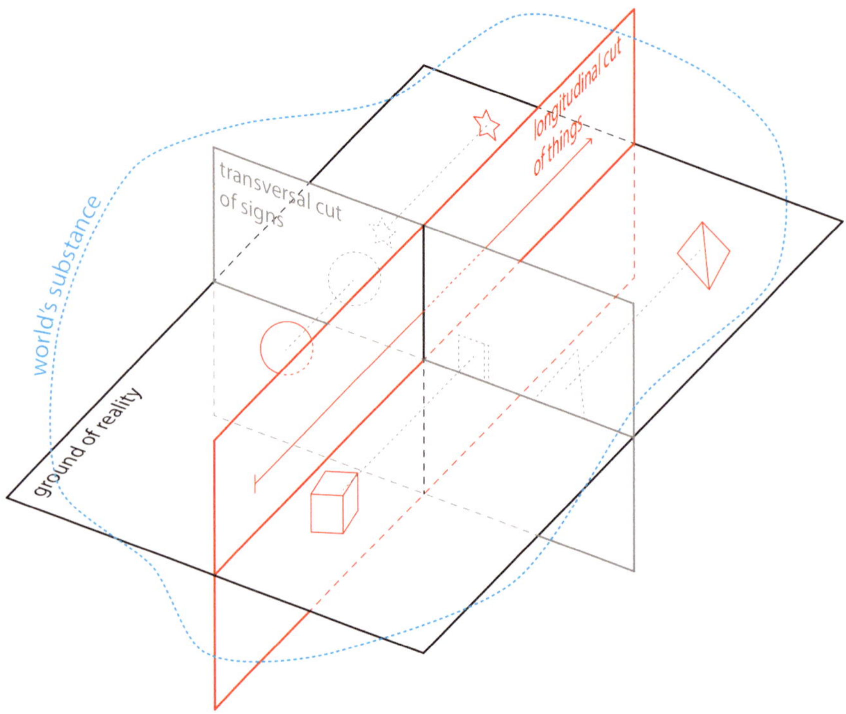

Fig. 2.2 Benjamin's model of the longitudinal cut of things and transversal cut of signs (*diagram* rp)

"if the vertical face is aligned with the human subject, then the horizontal landscape is the mode for all their processes of anti-Oedipalization: the 'body without organs', the 'becoming animal', the rhizome, nomadology, the war machine" (Hight 2003, p. 30).

With reference to Deleuze Guattari's resourcing to a 'masochism' for the conversion of a vertical organization, i.e., an Oedipal figuration or faciality, into a horizontal organization of intensities and desires, i.e., a 'body without organs' or landscape, Hight calls the rotation of axes that he describes a 'masochism upon a body of knowledge' (Hight 2003, p. 31).

After identifying architecture with the vertical position of the Oedipal subject, Hight concludes that "the proposition of landscape urbanism [...] attempts to rotate architecture out of its vertical alignment as a model of order, to deterritorialize [...] not the physical space of the city but the discipline's precepts and ethos". Landscape urbanism is, in fact, first of all a 'modality' and an 'attitude' (Hight 2003, p. 32).[26]

[26]At the same time, Hight also regards the prevalence of horizontal diagramming in landscape urbanism as a groundbreaking rotation of the traditional vertical alignment of landscape painting.

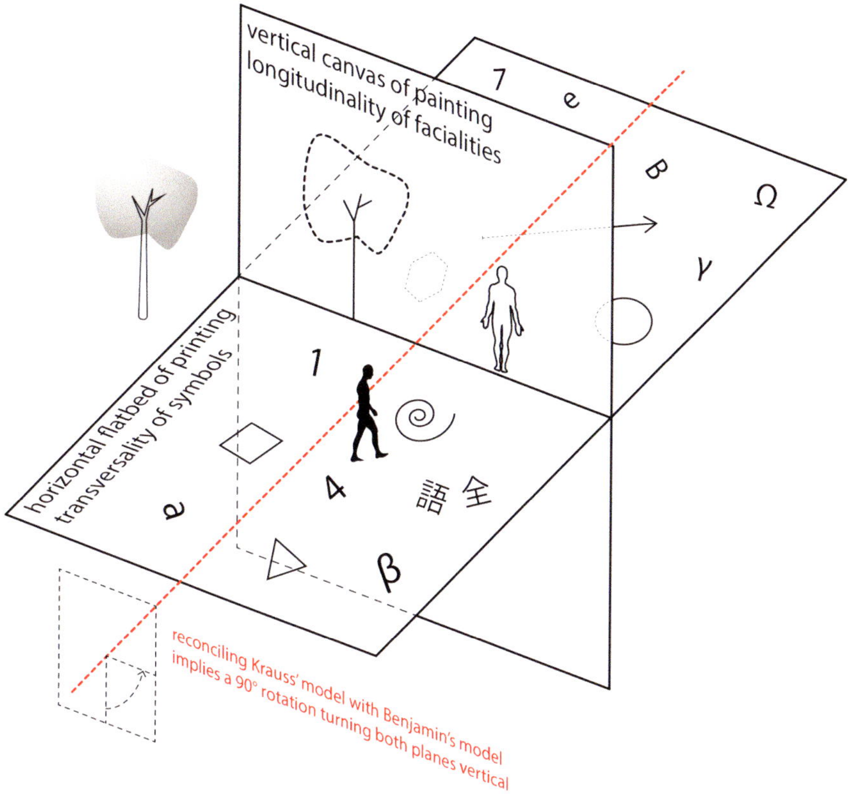

Fig. 2.3 Rosalind Krauss' model of painting's vertical canvas and graphic printing's horizontal flatbed (*diagram* rp)

2.5.5 *Planes Galore and Diverse Temporalities*

In truth, Deleuze Guattari's description of the processes of anti-Oedipalization in plateau 10 '1730: Becoming-Intense, Becoming-Animal, Becoming-Imperceptible…' of *A Thousand Plateaus* is probably the most adherent to the issues at stake and articulate among their various narratives where landscape comes into play.[27] Plateu 10,

---

[27]Plateau 10 seems more relevant to the matters of landscape than plateau 6 'November 28, 1947: How Do You Make Yourself a Body Without Organs?' quoted by Hight. In plateau 6, Deleuze Guattari are defining the term 'Body without Organ' as a Batesonian 'continuous region of intensity' or 'piece of immanence' (Deluze and Guattari 1987, p. 158) that overcomes Couvierian and structuralist taxonomies of the world. It is the narration of the term's essential, rather than spatial, characters that is the object of plateau 6. Plateau 10, instead, constitutes a much more substantial volume centered on the spatial exploration of natural unfolding. There, the sequence of Linnean serial, Couvierian/Levi-Straussian structuralist, and Batesonian immanent systems of classification of the spatial field is elaborated in greater operative detail (Deluze and Guattari 1987,

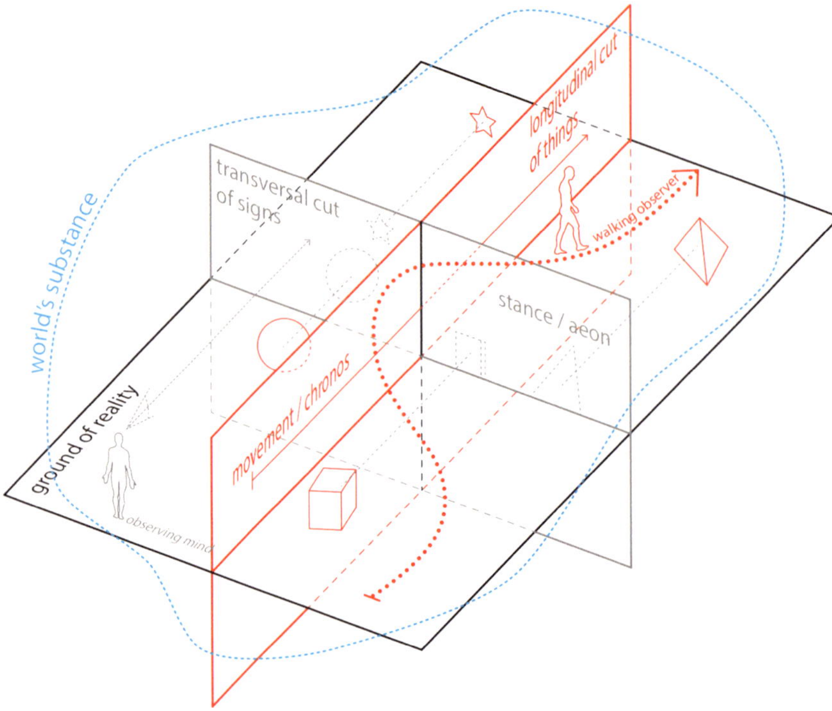

Fig. 2.4 An observing mind and a walking observer introduced to interpret Benjamin's model (*diagram* rp)

in fact, interfaces two plan(e)s that are then, to our understanding, both horizontal. Deleuze Guattari describe in fact the 'plan(e) of immanence' (introduced as 'plan(e) of consistency' and also referred to as 'plan(e) of haecceity' or 'the Octopus', i.e.,the 'Body without Organs') regulated by the non-pulse time of *Aeon* (that is geography), as coexisting with the 'plan(e) of transcendence' (the genetic plan(e) or teleological plan (e) or 'the plan(e) of subjectivities') regulated by the pulse time of *Chronos* (that is history) (Deluze and Guattari 1987, pp. 251, 254–5, 265–6). Although not explicitly stated, it seems quite ascertained all through the narrative that the construct envisions processes of 'becoming' converting the latter plan(e) into the former and vice versa in cycles of deterritorialization and reterritorialization. In any case, if the two plan(e)s are coexisting, they must be parallel and therefore both horizontal. In fact, the *incipit* of 'Memories of a Plan(e) Maker' reads "Perhaps there are two planes, or two ways of conceptualizing the plane" (Deluze and Guattari 1987, p. 265), prefiguring a planar coincidence.

pp. 236–7). Also, the compared anatomy of the Plane of Immanence and the Plane of Transcendence is explicitly formalized in sub-chapter 'Memories of a Plan(e) Maker' (Deluze and Guattari 1987, pp. 265–6).

The vertical alignment at stake in *A Thousand Plateaus* is relative to the 'subjectivities' or 'facialities' and their individual organizations, while their distribution remains over a horizontal plan(e).

2.5.6 Bartering Ethos

In architecture, the operational, organizational, metabolist, and functioning cut is, traditionally, that of the horizontal plan, way before that of the vertical section. After the plan, there comes the vertical silhouette of the façade, which is pictorial, picturesque, and aesthetic. Landscape painting, instead, projects the longitudinal organization of things onto the vertical canvas. In simpler words, far from registering a rotation of operative design planes, Hight's discourse is registering landscape urbanism's preference for horizontal symbolic diagramming over perspectival figurative representation and exploration of the landscape, that is landscape urbanism's adoption of architectural modalities in the practice of landscape. The 'ethos'[28] of architecture is in this case colonially (heteronomously) applied to the practice of landscape rather than the opposite. In Saidian terms, that corresponds to an action of reterritorialization by the colonizer on the colonized, prevailing over the specular deterritorialization of the colonizer produced by the exposition to the 'Orient'.[29] Indeed, it is more of an action of heteronomous colonization over the practice of landscape design, rather than an autonomous reformulation of architecture, despite secondary reciprocal contagions. To this respect, landscape urbanism turns landscape manipulation into a more profitable practice. Using Deleuze Guattari's phraseology, we could describe this passage in the practice of landscape design as "the serial theme of sacrifice [yielding to] the structural theme of the institution of the totem" (Deluze and Guattari 1987, p. 237): a figuration representing the action of modernization operated by Levi-Strauss tending to replace the analogy of series with the analogy of proportion in the interpretation of the world. Deleuze Guattari's BwO lies a step further.

2.5.7 Horizontal Organizations of Vertical Facialities

In an alternative interpretation, we propose that Benjamin's 'longitudinal cut of things' be the horizontal axis along which the observer slides ahead in its dynamic exploration of the assemblage (distribution) of subjective entities (vertical facialities) performed through a continuous visual capture. The longitudinal cut of the world is thus a diachronic (*Chronos*) and naturalistic cut. The 'transversal cut of

---

[28]See sub-chapter 'Masochist planes'.

[29]See sub-chapter 'The colonial perspective of landscape urbanism' in Chap. 5 of this book.

signs', instead, is the punctual perception, or mental section of reality, operated not by the sight, but rather by the mind of the immobile observer, nailed onto its vertical axis (*Aeon*). The transversal cut of the world is thus an instant and abstract cut. Once ascertained that in Benjamin's image both planes are vertical, we could even convene that the perception of the 'transversal cut' is prevalently vertical, while the observation of the 'longitudinal cut' is prevalently horizontal as it is performed through a trajectory sliding over the horizontal plane.

The 'longitudinal cut of things', no longer projected onto the picturesque canvas, is thus identified with the horizontal organization peopled with vertical facialities (Fig. 2.5).

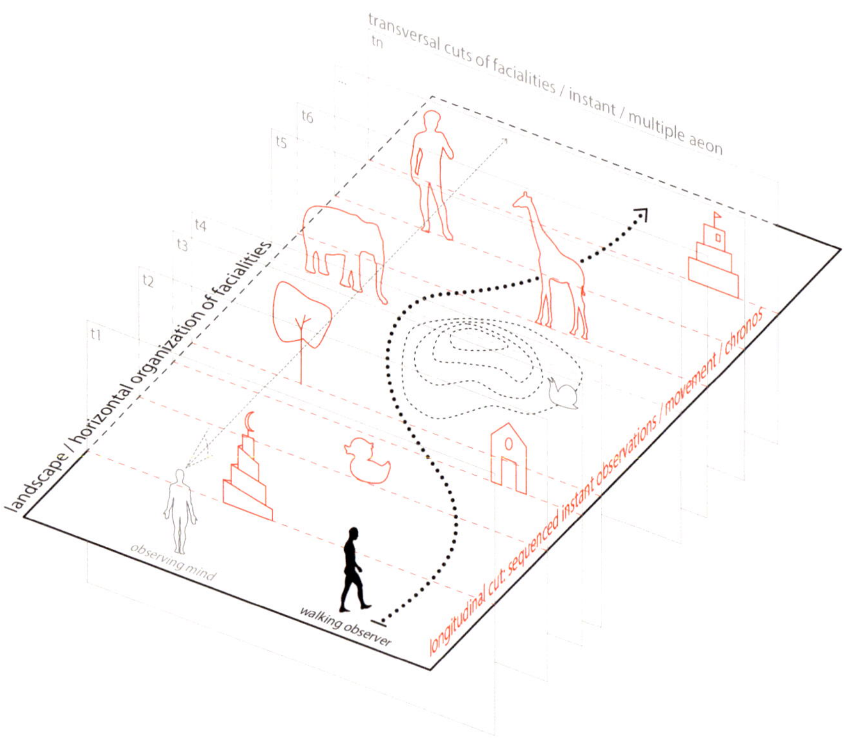

Fig. 2.5 Benjamin's model of the longitudinal cut of things and transversal cut of signs combined with Deleuze Guattari's description of the landscape as a horizontal organization of vertical facialities (*diagram* rp)

2.5.8 Multiplication of the Transversal Plane

The 'transversal cut of signs', instead, is encoded into the vertical stratification of multiple horizontal operative diagrams, where the interactions among different levels constitute the relational fabric of the landscape. The prevailing verticality, then, is not that of pictorial facialities, but rather the vertical stratification of horizontal operational diagrams that interpret the region. That vertical multiplication of planes interlocks both temporal states and thematic mappings.

Multiple transversal cuts register the vertical projection of the perspectival and subjective perception of the horizontal organization. The transversal plane is multiple since it is different for each subject and also since the perception of the same subject changes along the trajectory of exploration. A trajectory of exploration also implies a longitudinal timeline, but just a secondary subjective timeline in relation to the comprehensive timeline relative to the region. The longitudinal axis registers this subjective temporal unfolding, whose character is fundamentally figurative, by projecting the facialities of the region onto the picturesque canvas at discrete times (Fig. 2.6).

The horizontal plane registers, by means of symbolic diagramming, the operative organization of the region under various themes. The vertical multiplication of the horizontal plane registers the primary temporal unfolding of the region at given moments. This model of verticality is compatible with the construct of verticality we have described while discussing Marot's sub-urbanist proposal. George Descombes' project for the park of Lancy analyzed by Marot in 'Between Garden and Map' (Marot 2003) shows the potential of a 'verticalizing' design to become an *in situ* map: that is a multilayered cultural machine capable to rehabilitate a landscape as a repository of collective memory, while orchestrating its environmental re-engineering.

2.5.9 For a Technocratic Governance of the Environment

In *Landscape Infrastructure* (2012), Pierre Belanger traces the roots of a noble American lineage for the metabolist model of landscape urbanism. Lewis Mumford's identification of the 'foundation role of landscape' in shaping 'urban economies' implies in fact the centrality of a technocratic governance of the environment. Belanger refers to the "three main ways of modifying and humanizing the visible landscape: agriculture and horticulture; […] city development and architecture; […] works of engineering" (Belanger 2012, p. 310, note 1). However, the implementations of landscape urbanism have often conceded to fashionable formalisms, whereby conventional park designs deploy the abundant presence of vegetation within metropolitan contexts to veil hetero-directed operations.

Munford's work is a convenient base for both metabolist and culturalist elaborations, as for Rossi's monumental poetic construction dedicated to the city,

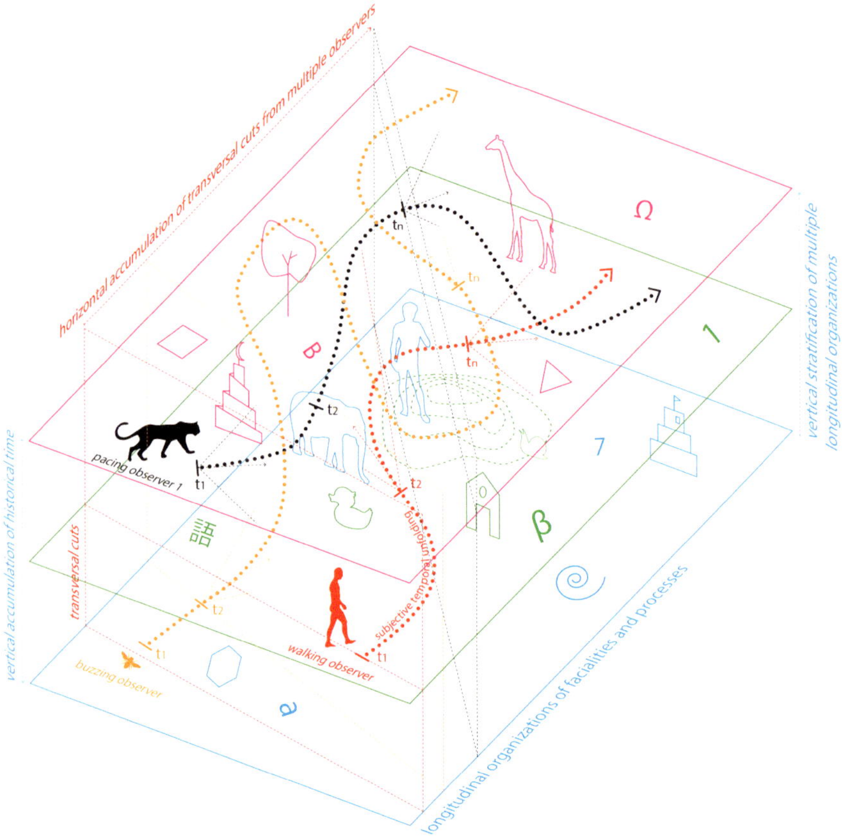

Fig. 2.6 Alternative landscape model implying the multiplication of the observers, the free rotation of the axis of the transversal planes, and the vertical accumulation of horizontal organizations of facialities and processes (*diagram* rp)

founded on the antipodal assertion that "the city is a fact in nature, [...] [but] with language itself, it remains man's greatest work of art"[30] (Rossi 1982, p. 180). Mumford's investigation of landscape formation, however, is bound as much to the environmentalist/machinic apparatus model as it is to a semanticizing model, evoked by Marot and Hunt on ancient bases, and to Patrick Geddes' anthropological 'Valley Section' fresco.[31] The 'Valley Section', in fact, is a militant

[30]As we mentioned before, Rossi is quoting Mumford's formulation along with Claude Lévi-Strauss' remarks about *la ville* as standing "at the point where nature and artifice meet". Rossi summarizes Lévi-Strauss' take as "an object of nature and a subject of culture".

[31]See the double illustration: 'The Valley Section and its social types: in their native habitat and in their parallel urban manifestations' (Geddes 1949b, pp. 166–7). See also the accompanying text: "This relief and contour is also associated with a kindred diagrammatic picture of the primitive occupations conditioned by this relief. This serves as an introduction to Rational Geography of

manifesto depicting the dwelling space of man, the landscape at large, as the result of the unfolding of collective sociocultural dynamics, rather than a technocratic initiative, over the topographic platform (Geddes 1949b, pp. 163–7).

2.6 Crossing Culturalist Terrains

2.6.1 Embedding Semantic: For an Aesthetic Mastery of the Surroundings

As Marot notes referring to Cicero via Hunt, in a semanticizing landscape model, "thought can lodge [in earth] its own signs and display the full range of its states and moods" (Marot 2003, p. 20), that is landscape as a manmade nature, rather than a matter of design, is a matter of embedding semantic, or language into the land.

Undoubtedly, the idea of embedding semantics into the land has ancient roots. In his *Greater Perfections*, Hunt reinterprets a letter of the sixteenth-century Lombard humanist Jacopo Bonfadio to retrace the origin of the landscape model based on the idea of a 'trio of natures'. By this model, the first nature is fundamentally a cultural perception of a pristine real or legendary wilderness, the second is the space transformed by man for productive purposes, the third is a space where a linguistic formulation, something we have referred to as mythology, is embedded, as in the garden (Hunt 2000, pp. 32–75). The first nature is born at once with the second when the agricultural turn spins the transition from the hunting-gathering to the herding-cultivating condition. The appearance of a manmade space illuminates the presence of the wilderness beyond its limits. Pristine nature is loaded with mythical values and can be colonized by means of hierographies to inscribe the 'numinous' in it (Hunt 2000, pp. 53–4, 61). The third nature descends from the demise of exclusively productive purposes in the manipulation of land. A contemplative attitude, allowed by the surplus accumulation of resources, seeks aesthetic and cultural pleasure in grace,

Cities, in terms of their regional origins. These are best studied and understood, to begin with, by beginning with the Valley Section and its resultant occupations and corresponding types of settlements. Note the Miner, the Woodman, and the Hunter on the heights; the Shepherd on the grassy slopes; the poor Peasant (of oats or rye) on the lower slopes; and the rich Peasant (with wheat, and in south it may be wine and oil) on the plain; finally, the Fisher (sailor, merchant, etc.) at sea-level. For thus it is that cities have arisen and still arise. As the merchant nobles of Venice sprang from the fishing-boat, or the millionaires of Pittsburg now arise beside the forge, so surely also do their cities retain the essential character, that conditioned by their environment and occupation." (Geddes 1949b, pp. 165–6). The text of the 1949 edition is from the catalog to Geddes' first "Cities and Town Planning Exhibition", London, Edinburg, Belfast, Dublin 1910–1911, while the drawings are from the second exhibition, Madras 1915, assembled in India after the sinking of the first exhibition *en route* to India.

that is the "Greater Perfections" of the garden, according to Francis Bacon's definition that gives the title to Hunt's essay (Hunt 2000, p. 63). [32]

The *incipit* for Hunt's discourse on the three natures is Bonfadio's epistle to Plinio Tomacello of 1541, describing his country retreat on Lake Garda. Hunt elaborates on the use of the expression '*terza natura*' (third nature) referring to the gardens where the '*industria de' paesani*' (the industry of the villagers) has 'incorporated', that is merged, nature and art (Hunt 2000, pp. 32–4). Analogies with Pliny's epistle to Domitius Apollinaris describing his *Tuscum* estate are noted by Hunt but not elaborated upon.

It results quite clear from the original text that, by 'third nature', Bonfadio plainly refers to a 'third thing' that is the combination of the two terms 'nature' and 'art'. However, upon what is little more than a pretext, Hunt builds his sophisticated formulation by tracing back wishful threads to Cicero's description of an '*alteram naturam*' ('other nature') in his *De natura deorum*. Cicero's '*alteram naturam*' is 'created' by man's hands '*in rerum natura*' ('within the natural world') by sowing crops, planting trees, redirecting rivers for irrigation purposes.[33] The replacement of the terms 'nature' and 'art' from Bonfadio's epistle with Cicero's 'natural world' and 'other nature' (for Hunt 'primal nature' and 'agricultural land'), allows the 'third nature' of Bonfadio's gardens on the lake to acquire a different sense and cast the prefiguration of the 'trio of natures', that is a 'cultural landscape' notion. The 'industry' that the Garda villagers infuse into primal nature to create the alternative nature of their gardens is the cultural agent that molds the landscape. The result is a material narrative, exceeding the productive purposes of agriculture, aimed to account for a community's origin and reason in the world, that is villagers' own material mythology.

Bonfadio's subsequent descriptions of the "world of citrus and olive groves, orchards and green pastures" on the shores of Lake Garda, in opposition to the surrounding "arduous [...] and menacing mountains" (Hunt 2000, p. 34), acquire a different projection in Hunt's discourse, composing the 'trio' of primal nature, agricultural land and garden into a sophisticated idea of landscape. Hunt knows

[32]Marot's references to Hunt are frequent and explicit. See his critical reading of Michael van der Gucht's etching, frontispiece to the *Curiositez de la nature et de l'art sur la vegetation* by the Pierre Le Lorraine Abbé de Vallemont in 'Envisioning Landscapes' on Harvard Design Magazine (Marot 2013, p. 92). The etching is described by Hunt as the explicit representation of his 'trio of nature' interpretative model of landscape-making (Hunt 2000, p. 40).

[33]Around 44 BC, Cicero states: "Terrenorum item commodorum omnis est in homine dominatus: nos campis, nos montibus fruimur, nostri sunt amnes, nostri lacus, nos fruges serimus, nos arbores; nos aquarum inductionibus terris fecunditatem damus, nos flumina arcemus, derigimus, avertimus; nostris denique manibus in rerum natura quasi alteram naturam efficere conamur" (Cicero, De natura deorum, par. 152). In English: "All the useful terrains, also, are mastered by man: we take advantage of the fields and mountains, ours are the rivers and lakes, we sow corn and we plant trees; we give fertility to the lands by channeling the waters, we coerce, divert, and redirect the rivers; by our hands, eventually, we almost attempt to create another nature within the natural world" [translation by the author].

well the Renaissance practice of summoning wishful authorities from the past to backup new ideas. He probably follows a similar pattern for the pleasure of a subtle intellectual construction.

2.6.2 *Of Lovely and Horrid Epiphanies*

The intense interest from famous or obscure humanists in the observation of the territory throughout the fifteenth and sixteenth centuries has been extensively explored by Piero Camporesi in *Le belle contrade* as the incubator of the modern notion of landscape (Camporesi 2016, p. 5).[34] In 'Vaghezze e orridità' (Lovely and Horrid Epiphanies), Camporesi discourses on the descriptions of the idyllic grooves on the shores of Lake Benaco (or Garda) from Ludovico Carbone, Pietro Cataneo, or Agostino Gallo, while lines from Gio. Francesco Tinto's *La nobiltà di Verona* of 1590 transfigure the incumbent horrors of Mount Baldo's peaks into a shrine of herbal treasures, whose deep valleys are frequented by 'physicians, surgeons, and apothecaries' from all over (Camporesi 2016, p. 143 and p. 198, note 21). The 'precious *hortus sanitatis*' of Mount Baldo bears the signs of a slow but incipient transition towards a sentiment of sublime, whose aesthetics will be formulated by Edmund Burke almost two centuries later (Camporesi 2016, p. 148).

In 'Paesaggio e lavoro umano' (Landscape and Human Action), Camporesi reconstruct the novel gaze shed over the lands in the mid-sixteenth century to acknowledge man's modifying action on nature through a myriad of literary fragments. We learn how, before, the late capital of the Western Roman Empire, Ravenna, had become a "fragment of the Orient plunged into the marshes", "Byzantium on the hazy Adriatic", "aborted prototype of Venice", "paradoxical city", "inverted world" (Camporesi 2016, p. 165),[35] and its lands "*cloaca foetidissima Galliae Cisalpinae*" (Cisalpine Gallia's filthy cloaca), where "*sitiunt vivi, natant sepulti*" (the living thirst, the dead float) (Camporesi 2016, p. 165).[36] It is the Dominican preacher Leandro Alberti that first "reconnoiters the real country" following the "long hydraulic strive" that had reclaimed the lands of the Bassa Romagna into the "very fertile plains" later to be noticed along Michel de Montaigne's precursory Grand Tour in 1580 (Camporesi 2016, p. 164). According to Camporesi, through his descriptions of those lands, Friar Alberti recognizes the confluence of the "history of the work of man", his "industrious knowledge" and the "impassible power of nature" in the notion of "territory" that he deploys largely in his descriptions of the Italic peninsula. As in his description of the agrarian

---

[34]In Chap. 1 (Introduction) of this book, we have argued for the translation of the Italian term '*contrade*' to English as 'shires' for its capacity to evoke the transition from a medieval to a modern thought on the landscape.

[35]Translations by the author from the Italian original.

[36]Camporesi's first quote is from Giovanni Boccaccio's letter to Francesco Petrarca, the second quote from C. Sollii Apollinaris Sidonii's *Opera*.

landscape of the Campania Felix around 1550, where "art, that is intelligence, ability to devise as to implement" and the "powerful nature" have "collaborated to the construction of an admirable 'earthly paradise' (from Alberti's *Descrittione di tutta Italia*), an enormous outdoor villa" (Camporesi 2016, pp. 166–7).

The same *amoenitas* (beauty) descending from the "domestication of wilderness" is to be found in the "sharp visions of a laborious landscape" from Andrea Bacci, archiater to Pius V, distinguished scholar of thermal waters, of Tiber, of natural history, and, most of all, of the "history of the cultivation of grapevine and wine, of which civil history has been said to be an appendix". As in his description of the *ager* of Cesena around 1595, where the perfect organization of olive groves, orchards, and vines in *quinconce* according to the precepts from Columella, and some more ancient pre-Etruscan practices, 'overtakes any other most exquisite spectacle of the countryside' (from Bacci's *De naturali vinorum historia*) (Camporesi 2016, p. 168).

With "the scent of the myrtle, laurel, jasmine, roses, and rosemary", Camporesi's recount of the '*belle contrade*' absorbs from his sources the physical and human geography of the Quattrocento and Cinquecento, as to be neither a historical, nor a critical analysis, but rather a material evocation in the space of literature.[37]

2.7 A Second Critique of Metropolitanism as Urban Age

2.7.1 The Urban Age

According to Ricky Burdett, the 2004–06 Urban Age project started to debate the relationship between physical form and social well-being, at the turn of a century in which 'world's population living in cities' passed from 10 to 50% of the total, a share projected to raise rapidly to 75% by 2050 (Burdett and Rode 2007, p. 8). Burdett's narrative universally identifies the physical form of generically anthropogenic space with the urban form, while the social referent remains specific, that is the urban society, by that generating a deceiving parallel: the identification of the metropolitan with the urban.

The project strives to bring together four technocratic categories of agents, active in the process of transformation of the city, namely city leaders, policy-makers, design professionals and academics, building a formal network. Across eight itinerant conferences the project focused on the six global- or mega-cities of New York City, London, Berlin, Shanghai, Mexico City, and Johannesburg, showing an idiosyncratic selection of cases that underrepresents South- America and the Mediterranean, probably due to utilitarian tactics.

[37]All quotations from Camporesi in this sub-chapter are English translations by the author from the Italian text.

The representation of the chaotic expansion of the Latin-American global- or mega-city is delegated to Mexico City anomalously sitting in North -America. The outdated case of Curitiba's BRT bus network recurs, along with Bogotá's Transmilenio bus, as a constant reference for "less-onerous" (Burdett and Rode 2007, p. 22) solutions of public transport in developing regions. The bus system's inefficiency and the consequent spatial fragmentation and social seclusion are disregarded in the construction of a false myth that substantially contributes to the edification of a positive image of the global city outside of the developed world. The flat mystification stating that "Bogotá's bus lanes represent an ideal of easy movement for the masses" (Sudjic 2007, p. 44) seems to derive from a naïve postcolonial perspective in tune with the immediate interests of the agencies financing the project.

The Urban Age project celebrates at once the apotheosis and the epitaph of metropolitanism, while smuggling its exclusive narrative as a promise of democratic urbanity.

2.7.2 *Total Khéops: la misère est moins dure au soleil*

Moreover, the unresolved expansion of vast peripheries around historical centers still bound to the suffering but resilient urban tradition of the Mediterranean, is disregarded in the study. The plotting of the system of the 'opposed but accessible shores' originated from the coastal centers, proposed by Hashim Sarkis as a region-forming process in accordance with David Abulafia's 'human history of the Mediterranean', will be analyzed more in detail later.

The urban landscapes of Jean-Claude Izzo's *Total Khéops* render a glimpse into the deep complexity of the theme (Izzo 1995). By describing the existence of a Mediterranean hinterland of timeless poverty beyond the globalized periphery of recent marginalization of Marseille, Izzo delineates two distinct and hardly communicable spheres of poverty and marginality grounded in the region, one mitigated by the sun, the other not.

Fabio Montale, the '*macaroni*', son of Italian immigrants become policeman, lives in a little house, the miserable inheritance from his parents, just a cabin of bricks, planks, and tiles built on the rocks of the shore of Les Goudes, South of the city. Across the sheer cliffs of Les Calanques, Montale treks down a short trail to see again the fishing boat at the moonlight. He knows that the *bouillabaisse*[38] is waiting in a pot prepared by Honorine the old fisherman's widow.

[38]The *bouillabaisse*, *bolhabaissa*, *bugliabasciu*, κακαβιά, or *kakavia*, contains the whole of the Mediterranean, its fish and its souls, as an *hortus conclusus* mirrors the totality of the universe. Greek colons of Phocea, fleeing civil conflicts from the coast of Asia-Minor, came to found Marseille on the coast of present-day Provence between France and Italy. Their fishermen probably first prepared the *kakavia* in the sixth-century BC using the fish that could not be sold on the market. The *bouillabaisse* has Greek, Middle-Eastern, Franc, and Italic roots. The *bouillabaisse* can span from basic to sophisticated preparations without losing its identity.

Up North, in building B7 of the modern project of Cité La Paternelle, a family of North-African immigrants is dispossessed of their Mediterranean identity and Mouloud's young daughter Leila is handed-into an ominous destiny to be annihilated. At Cité La Paternelle

> Not a tree, nothing. The housing. The parking. The *terrain vague*. And in the distance, the sea. L'Estaque and its port. Like another continent. I remembered that Aznavour used to sing: *La misère est moins dure au soleil*. No doubt he had never come here. Up to these clumps of shit and concrete (Izzo 1995, p. 22).[39]

2.7.3 *Massif Les Calanques and Concrete Plateau*

The topography of the massif of Les Calanques is opposed to the concrete slab of the project's plateau: the *tearrain vague* of the *banlieue*[40] is an incommensurable 'continent' to the terrain of poverty of Les Goudes, as it is to L'Estaque. Spaces made of a different stuff, or quality. The immigration waves of Italian '*macaronis*' fleeing fascist purges, along with Greek and Spanish republicans escaping francoist dictatorship, remapped their topographies over the terrains of the Panier and Vieux Port in the ancient center, as well as in the *anciennes hameaux* along the seashore, from Les Goudes to L'Estaque. It is the miserable accumulations of Mediterranean urbanity that receives the lives of these Southern European refugees, while the Maghrebi clouds of outcasts and unprotected youth are alienated over the concrete plateau of the post-war era. In *Histoire universelle de Marseille*, Alèssi Dell'Umbria reconstructs the metamorphosis that converted the *anciennes hameaux* from fishermen villages to *villages d'ouvriers* at the end of the sixteenth century and then their touristic gentrification by the beginning of the twenty-first century (Dell'Umbria 2006). By reinventing the psycho-geography of Marseille, Izzo's *trilogie marseillaise* contributes to grounding the physical geography of the northern shore of the Mediterranean at the end of the twentieth century.

The surviving nuclei of the production of urban space in the Mediterranean, and their resilient ambits, resisting expansion, tourism, and gentrification, exact a specific consideration in the face of the later production of globalized metropolitan space. As well as exact specific treatment the urban myths, deeply ingrained in the history of urbanity, as for the *locus* of Rome or Paris, disregarded by the metropolitan model of urbanity.

---

[39]Translation by the author from the French text.

[40]By *terrain vague*, Izzo does not refer to Ignasi de Solà-Morales' notion, claiming to the metropolitan spaces of abandonment aspirations that anticipate by a decade Gilles Clement's *tiers paysage*. He rather refers, here, to a more general meaning of the French expression associated to the *banlieue*, to the very absence of the city in its terrain rather than to its absence as a space of possible form. See *Presente y futuros. Arquitectura en las grandes ciudades* (de Solà-Morales 1996).

2.7.4 Pick Your Urban Age à la carte

The Urban Age project analyzes six models of global city, through the lens of five 'deeply-connected' binaries that matter to the "environmental, economic and social sustainability of global society": 'social cohesion/built form', 'sustainability/density', 'public transport/social justice', 'public space/tolerance', 'good governance/good city' (Burdett and Rode 2007, p. 23).

Mexico City "epitomizes the tensions between spatial and social order" of a land-consuming endless expansion whose recipe lies in policies for region-wide growth containment, redensification of the consolidated center, rail-based public transport (Burdett and Rode 2007, p. 11).

Johannesburg's mosaic of "walled shopping centers and gated residential communities" seceded from its demised down-town composed of the urban ruins of dilapidated condominiums turned into makeshift-kitchen apartments, is a "physical landscape that monumentalizes separation over inclusion" (Burdett and Rode 2007, pp. 11–2). For "a place where public space fails to perform its democratic potential as a place of interaction and tolerance" the recipe is to prioritize public transport, invest in retrofitting the center's social spaces and facilities, and contain suburban expansion (Burdett and Rode 2007, p. 18).

New York core's status of minority-majority melting-pot city testifies about the capacity of "built form in sustaining cycles of urban change" in conjunction with an efficient metro network. The recipe is the strategic coordination of its fragmented governance to retrofit the brown fields of the "derelict industrial sites" surrounding Manhattan into a "blue belt of linear parks and open spaces" (Burdett and Rode 2007, p. 18).

The "heroic scale of pace and change" leading the "Shanghai urban experiment" accumulated 200 towers per year for 25 years in the pursuit of economic progress, on the one hand doubling the per capita floor area in fifteen years under the "overpowering demand" from emerging middle classes, on the other hand at cost of the "forced relocation of inner city dwellers to remote high-rise estates", had a radical impact on the public realm at ground level, producing a terrain of "isolated point blocks" surrounded by motorized mobility infrastructures (Burdett and Rode 2007, p. 19).

London's early-2000s challenge to "accommodate all growth within the city's existing boundaries" was centered on the rehabilitation of the semi-central brownfields of the docks introducing high-rise constructions dedicated to business and mobility hubs combined with cultural programs of excellence. The rehabilitation was mainly linked to large public investments for the extension of underground Jubilee Line favoring financial speculation of private groups. In spite of the 50% of housing quotas reserved for social housing, processes of gentrification can be recognized in the transformation of the inner areas of a city where 50% still live in poverty (Burdett and Rode 2007, pp. 19–20).

"Poor but sexy" Berlin attracts "young, adventurous and bohemian" new residents by capitalizing on both the "all-encompassing lustre of [...] the reborn center on both sides of the former Wall" and the consequent stagnant economy produced by the money "run out for the museums, opera houses, and theaters", an unusual but consequent combination. Super-efficient public transport on rail, lavish equipment of parks and green areas, high spatial quality of the urban form and correspondent maintenance, coupled with a regime of affordable rents in the absence of corporate economy, favor a high birth rate, café culture, promiscuous night life (Burdett and Rode 2007, pp. 21–2).

2.7.5 *Globalization, cet obscur objet du désir*

Across the review of the case studies, the dialectic among the five mentioned binaries of 'cohesion/form', 'sustainability/density', 'transport/justice', 'space/tolerance', 'governance/city' could be reduced to a confrontation between a global North and a global South, at least strictly in terms of urban form. In fact, the "resilient urban structures" of New York City's grid with mixed-use multi-story buildings, London's terraced houses, and Berlin's perimeter blocks, capable to absorb large part of the effects of globalization behind their "active street frontages", are confronted with the sudden and dramatic shifts affecting the physical as well as the intangible contexts of the world cities of Shanghai, Mexico City, Johannesburg. Globalization, described as a phenomenon with "positive impact on local economic development", but "negative physical effects on income disparity, social exclusion, and an increasingly ghettoizing landscape", remains indeed the occulted focus and *cet obscur objet du désir* of the Urban Age project (Burdett and Rode 2007, p. 22).

2.7.6 *Anachronism*

The Urban Age agenda that descends from the study results even more simplified, summarized in a "compact, mixed-use, well-connected, complex, and democratic city" taking up the challenges of "globalization, immigration, jobs, social exclusion, sustainability" to turn them into opportunities (Burdett and Rode 2007, p. 22). It is, in fact, only one model of global city that is promoted by the Urban Age corporation, to which every city on earth is called to homologate. It is in fact a model of global city, or world city, or Cosmopolis, as opposed to the 'city-world',[41] functional to the cultural as well as economic forces of globalization, characterized by the univocality of the process of conformation of the global South to the global North, excluding the possibility of any miscegenation.

[41]See Chap. 3, sub-chapter 'Making the world' in this book.

Realistically the LSE Urban Age project elaborated between 2004 and 2006 and published in 2007, is an optimistic picture of the big city definitely bound to the time before the 2008 recession, a vision that now, after only few years, results anachronistic.

The projected image indulges the false myth of globalization seen as a phenomenon of economic and material progress that can be canalized and controlled in order to diffusively take advantage of the possible economic growth, equally in the congested environment of Manhattan or the metrosexual borough of Brooklyn and in the segregated townships of Soweto and Alexandra. The negative impacts of globalization on communities and their landscapes, such as economic, social, and spatial fragmentation, are described as amendable, if not collateral, side-effects, controllable by means of a smart strategic coordination of technocratic élite summoned by the project, composed by the political, administrative, professional, and academic corporations active in the city-making process.

2.7.7 *Dioscuri of Globalization*

'Dissensus' to globalization is flogged by obloquy, as Sudjic states that "[...] it is possible to see the whole anti-globalization narrative, with its curiously irrational distaste for innocent companies such as Starbucks (who, after all, dispense coffee rather than napalm), as a contemporary twist on the same xenophobic themes that have made rural peasant communities suspicious of big-city ways throughout the centuries" (Sudjic 2007, p. 48). Furthermore, in Burdett's essay, the blatant process of conglomerate centralization in finance as well as all sectors of industry is contrasted with the phenomenon of diffusive networking of the world via information systems, occurring at an equally vertiginous and even disruptive pace. Although, it is certainly open to debate whether the latter corresponds to a real form of democratic empowerment of people or to a narcotizing mass-delusion complementary to the former.

While the false myth of metropolitan urbanity is built upon the narrative of globalization, it is also dependent on the apparent opposite narrative of happy informality. As a mondialist rhetoric, the latter is usually narrated by people comfortably living in the first world or cruising the globalized skies depicted in each day more recurrent maps of air-travel global connections as part of the contemporary jet-élite of fast mobility.[42]

[42]See for example (Burdett and Rode 2007, pp. 30–1). Metropolitan urbanity and happy informality are novel Dioscuri, reciprocally bound like the half-brothers Pollux, divine offspring of Zeus, and Castor, son to cheated mortal Tyndareus.

2.7.8 The Other Half

The disregarded half of the London School of Economics and Political Sciences, which does not find hospitality in the LSE acronym, proposes a radically alternative narrative of the urban phenomenon in the age of globalization. Consistent research work has been consolidated in the last 20 years by Sunil Kumar on the problem of housing production for the unprotected masses of the urban poor in the global South (Kumar et al. 2001) and the surprisingly non-conflictive dynamics of public policies and speculation and rental markets (Kumar 2008). Kumar explores the housing problem in the global South in its tight links to the issues of labor market exploitation, often disregarded internal migrations, and high fragmentation within the very scenarios of inequality (Kumar and Fernández 2015). Kumar asserts that the programs of social mobility tending to emancipate the lowest strata of urban society do not bring definitive solutions to the emergencies of the global South, but rather generate a vacuum that is going to be soon filled up by domestic and foreign migration of new groups, attracted to the periphery of the global city in search for economic amelioration. Parallel policies should tend to elevate the economic and, generalizing, living conditions as well as the dignity associated to jobs, functions, activities of the lowest and most unprotected sectors of the labor market. The acknowledgement of the unavoidable permanence of the unglamorous monolith, onto which the flamboyant crystal of metropolis is edified, is constantly overshadowed in the optimistic fresco propounded by the urban age adepts. While Koolhaas coaxes the masses with the elitist narrative of the advent of the 'Metropolitanites', the Urban Age project clouds the impossibility of the universal emancipation under the conditions of the global city system.

2.8 Chapter's Conclusions: A Novel Cityness

2.8.1 Cityness

At the demise of the century-old nation states as protagonist of the world economy, the LSE Urban Project identifies the new leading forces with the primordial entities of the millennium-old cities, although on an expanded scale. By opposing the reading of a dramatic change in the character of the contemporary settlements of geographic scale, under the various names of mega-regions or agglomerations, a substantial continuity of the urban space is advocated. While dismissing the possibility of a "post-urban form", that is a formulation accounting for the alterity of recent spatial organizations, Sudjic claims that only the traditional attributes of urbanity, such as "pedestrian public spaces" and "casual interaction between strangers" discern what can be called 'cityness'. Those principles must be reinjected into the expanded settlements, in order for those environments to be able to offer "a menu of shared experiences" (Sudjic 2007, p. 51). The implicit dispute here juxtaposes the idea of the city as permanent center of economy to Florida's idea of diffusion of the leading forces of the world economy over mega-regions, novel habitat of a creative class producing scientific innovation and its patented applications.[43]

2.8.2 *Geopolitical Post-metropolis*

While recollecting the results of the Urban Age project, Edward Soja and Miguel Kanai mention Florida's work explicitly (Soja and Kanai 2007, p. 61), considering him a later epigone in the wake of earlier territorial models, such as Geddes' 'conurbation', 'world city', and 'city region' of 1915,[44] Jean Gottmann's 'megalopolis' of 1961, and Constantinos Doxiadis' 'eperopolis' of 1968. However, according to Soja and Kanai, Florida highlight a primarily economic transformation that destabilizes the boundaries of traditional national economies by decentralizing consolidated industrial geographies (Soja and Kanai 2007, pp. 63–4). Cutting out a more critical reading of globalization in the context of the generally complacent Urban Age project, Soja and Kanai describe a sequence of phases of globalizing capitalism that go from the 'commercial capitalism' that created "mercantile world cities such as Amsterdam and London", 'financial capitalism' spreading worldwide with imperial colonialism establishing its 'global command posts' in London, Paris, and New York. The sequence is completed by today's information technology revolution originating the latest metamorphosis of a 'post-Fordist flexible capitalism'. The picture shows the novel centrality of 'information-based creative industries' as well as a true globalization of productive capital and the consequent formation of a "new and different global geography of economic development" (Soja and Kanai 2007, p. 63). Soja and Kanai describe the replacement of a 'First, Second, Third World' development paradigm with a new 'Global Division of Labour', tripartite in North America, Western Europe, and East Asia, as the wave of expanded 'techno-poles' causing the territorial imprinting of the novel mega-regional geography of 'metropolitan industrialism'.

Thus, Soja and Kanai regard the phenomenon of 'urbanization of the world', that gives the title to their essay as a process moving geography from more material to less material structures. On the one hand, Burdett clings to the attributes of urbanity,

---

[43]See tables ranking mega-regions according to the number of scientific citations and technological patents (Florida 2002, pp. 30–31).

[44]In 1915, Patrick Geddes had described the formation of regional settlements that he calls 'city regions' in the chapter 'World Cities and City Regions' of his *Cities in Evolution* (Geddes 1949a, pp. 22–31).

or the enhanced urbanity of metropolitanism, the '*cité à la carte*', to unfold the scenarios of a merry 'urban age', relying on an expanded version of the traditional attributes of the public space. On the other hand, Soja and Kanai describe a metamorphic state of spatial order, where the urban/suburban divide fades (called 'exopolis', meaning 'outside of the perimeter'), the 'urbanization of the world' corresponds to the 'globalization of the urban', and the configuration on the ground corresponds to the restructuration of the social order.

Despite Soja and Kanai's adhesion to the Urban Age mission expressed through their refusal of 'post-urban' formulations,[45] all the above calls for a set of different attributes to discern 'novel urbanity', or 'novel cityness', that overthrow the paradigms of 'new urbanist' and 'metropolitanist' coexisting in Burdett and Sudjic' formulations.

2.8.3 Metropolitan Demise: Retrofitting, Region-Forming, Physio-digital Responsiveness

Scenarios related to the demise of the exhausted 'metropolitanist' agency are being widely pursued. Authoritative academic institutions have cast varied horizons of reform of metropolitanism. Among other proposals: the convergence of all planning and design efforts towards the systematic retrofitting of existing structures and infrastructures on the territorial scale; the ecological retrofitting of informal settlements within never-completed vertical ruins of the metropolitan promise; the overcoming of "two decades of seeing architecture and urbanism as the spatial manifestation of the effects of globalization", focusing on the "emergence of the geographic" as today's "expanded agency of the designer" and regarding contemporary geographic configurations as traces of the action of anthropic forces; 'full scale prototyping' of systems 'responsive' to both 'political and cultural conditions' as well as fluctuating 'environmental factors', an attempt at bridging physical and digital spheres by "mak[ing] visible the invisible forces that shape our world" and architecture's potential to respond in real time.[46]

The process of opening to the geographic dimension, in fact, has crossed the field of design and overridden its agenda. A general reflection on the relationship between man and nature has escalated the list of priorities of the debate on design, pivoting on the issue of the transforming notion of landscape in the post-urban age.

[45]"Globalization and the formation of a New Economy have not been leading to a post-industrial or post-urban era, as many have claimed, but rather to a new and different round of urban industrialization that, in turn, is creating a new and different global geography of economic development" (Soja and Kanai 2007, p. 61).

[46]Respectively from: MIT's Center for Advanced Urbanism, ETH's Urban Think Tank, Harvard GSD's New Geographies research, Columbia's Living Architecture Lab.

2.8.4 *Semiotics of the Vicarious Encounter*

That of the possibility of a novel 'cityness' and its reformulated attributes in the expanded post-urban space is a crucial issue. To this respect, Rowe refers to a deeper semiotic dimension of man's inhabitation of space to discern the fundamental attributes of its hospitability. He appeals to the possibility of what he calls "vicarious encounter with special places that cause us to pause and truly reflect", that "certain depth and significance" (Rowe 1991, p. 59) provided to our life by the townscape. The semiotic layer, which Rowe calls the 'myth', marks in fact a profound divide between the idea of the 'middle landscape' and metropolitanism, new urbanism, sub-urbanism, on which we will elaborate in the next chapter.

References

Ábalos I, Herreros J (1995) Toyo Ito: light time. El Croquis, Toyo Ito 1986 1995 (71)
Alexander C (1977) A pattern language. Oxford University Press, Oxford
Arendt H (1998) The human condition. The University of Chicago Press, Chicago [1958]
Belanger P (2012) Landscape infrastructure: urbanism beyond engineering. In: In-frastructure sustainability & design. Routledge, London
Berger J (1972) Ways of seeing. BBC and Penguin Books, London
Buchli V (1999) An archeology of socialism. Berg, Oxford
Burdett R, Rode P (2007) The urban age project. In: Sudjic D, Burdett R (eds) The endless city. The urban age project by the London School of Economics and the Deutsche Bank's Alfred Herrhausen Society. Phaidon Press, UK
Camporesi P (2016) Le belle contrade. Nascita del paesaggio italiano. Il Saggiatore, Milano [1992]
Cavell R (2002) McLuhan in space: a cultural geography. University of Toronto Press, Toronto
Cicero MT (1933) De natura deorum, vol II, paragraph 152. In Cicero, book XIX (trans) Rackham MA. Harvard University Press, Cambridge. [44 BC]
Clément G (2015) "The Planetary Garden" and other writings (trans: Morris S). University of Pennsylvania Press, Philadelphia
Corner J (1999) Recovering landscape. Princeton Architectural Press, New York
De Feo V (1963) URSS Architettura 1917–1936. Officina, Roma
de Solà-Morales I (1996) Presente y futuros. La arquitectura en las ciudades. In: Presente y futuros. Arquitectura en las grandes ciudades, by AA.VV. Collegi Oficial d'Arquitectes de Catalunya/ Centre de Cultura Contemporània, pp 10–23
Dell'Umbria A (2006) Histoire universelle de Marseille: De l'an mil à l'an deux mille. Agone, Marseille
Deleuze G, Guattari F (1987) A thousand plateaus: capitalism and schizophrenia (trans: Massumi B). The University of Minnesota Press, Minneapolis
Farina A (2009) Ecology, cognition and landscape. Springer, Netherlands
Florida R (2002) The rise of the creative class. Basic Books, Philadelphia
Florida R et al (2007) The rise of the mega region. Research paper, The Martin Prosperity Institute, University of Toronto Rotman School of Management
Geddes P (1949a) Cities in evolution. Williams and Norgate Ltd., London [1915]
Geddes P (1949b) The valley section from hills to sea. In Cities in evolution. Williams and Norgate Ltd., London [1910–1915]
Guattari F (2000) The three ecologies (trans: Suton P, Pindar I). The Athlone Press [1989]

Halbwachs M (1909) Les expropriations et le prix des terrains à Paris (1860–1900). E. Cornely & Cie, Paris
Halbwachs M (1920) Les plans d'extension et d'aménagement de Paris avant le XIXe siècle Paris. Electronic edition by Tremblay J-M (2001). La vie urbaine, pp 3–28
Hight C (2003) Portraying the urban landscape: landscape in architectural criticism and theory, 1960–Present. In Mostafavi M, Najle C (eds) Landscape urbanism: a manual for the machinic landscape. Architectural Association, London
Hunt JD (2000) The idea of a garden and the three natures. In: Greater perfections. The practice of garden theory. University of Pennsylvania Press, Philadelphia
Ito T (1995) Architecture in a simulated city. El Croquis, Toyo Ito 1986 1995 (71)
Izzo J-C (1995) Total Khéops. Gallimard, Paris
Jünger E (2006) Il filosofo e l'Anarca. Intervista a Ernst Jünger. In: Volpi F, Gnoli A (eds) L'ultimo sciamano. Bompiani, Milano
Koolhaas R (1978) Delirious New York. Monacelli Press, New York
Kumar S (2008) Urban housing policy and practice in the developing world. In: Colby I, Sowers K, Dulmus C (eds) Comprehensive handbook of social work and social welfare: social policy and policy practice, vol 4. Wiley, Indianapolis, pp 249–294
Kumar S, Fernández M (2015) Urbanisation-construction-migration nexus/5 cities/South Asia. Research paper, LSE Enterprise, London
Kumar S et al (2001) Social relations, rental housing markets and the poor in urban India. Research paper, Department of Social Policy, LSE, London
Leccese M, McCormick K (eds) (2000) Charter of the new urbanism. McGraw-Hill, New York
Lévi-Strauss C (2012) Tristes Tropiques (trans: Weightman J, Weightman D). Penguin Books, New York [1955]
Magnaghi A (2010) Il progetto locale. Verso la coscienza di luogo. Bollati Boringhieri, Torino
Marot S (2003) Sub-urbanism and the art of memory. Architectural Association, London
Marot S (2013) Envisioning landscapes. Harvard Design Magazine (36)
Mostafavi M, Najle C (eds) (2003) Landscape urbanism: a manual for the machinic landscape. Architectural Association, London
Mumford L (1938) The culture of cities. Harcourt, Brace and company, New York
Rossi A (1982) The architecture of the city. The MIT Press, Cambridge [1966]
Rowe P (1991) Making a middle landscape. The MIT Press, Cambridge
Senkevitch A Jr (1982) Introduction. In: Ginzburg M (ed) Style and epoch. The MIT Press, Cambridge [1924]
Soja E, Kanai M (2007) The urbanization of the world. In: Burdett R, Sudjic D (eds) The endless city. The urban age project by the London School of Economics and the Deutsche Bank's Alfred Herrhausen Society. Phaidon, London
Sudjic D (2007) Theory, policy and practice. In: Sudjic D, Burdett R (eds) The endless city. The urban age project by the London School of Economics and the Deutsche Bank's Alfred Herrhausen Society. Phaidon, London
Yaro R (2000) Three views of the Northern New Jersey Region. In: Charter of the new urbanism. McGraw-Hill, New York

Chapter 3
The Geographic Prospects of Human Habitat and the Attributes of a Novel Urbanity

Abstract This chapter describes the expansion of the human settlements at the scale of mega regions and the consequent dissolution of the city/country division. The text also analyzes the concurrent expansion of an idea of city over larger territories, specular to the retreat of the relevance of urban space in such expanded contexts. The text discusses the emergence of a novel form of diffuse urbanity, articulating the synthetic considerations in the previous chapter. The spatial production overflowing the physical perimeter of the urban walls and, beyond those walls, trespassing the limits of a space commensurable with that of the urban paradigm, generates a distinct environment. The idea of a set of novel attributes defining the new form of citizenship is outlined, opening a vast field of future research work. The potential for the formation of a novel semantic and mythology, to endow the expanded territories with a new sense, is debated through the compared analysis of a set of spatial paradigms, such as the world city, city world, cosmopolis, and postmetropolis. Among these, particular attention is reserved to the model of the 'opposite but accessible shores' and its world-making ability. The bearing platform adopted for the compared analysis is constituted by a free reading of Rowe's 'middle landscape' construct, and its evolutions, such as the 'emergent architectural territories' that have been rising in East Asia, interpreted as a humongous experiment for the implantation of a fabricated mythology for a novel citizenship.

Keywords Mega region · Novel urbanity · Urbanity attributes
Middle landscape · Territorial effect · Beijing · Shanghai · Territorial mythology · City world · Cosmopolis

3.1 Introduction

A varied set of contemporary territorial conditions can be inferred from the observation of the mapping of present-day population diffusion across the planisphere: large regions inhabited at high density rates spreading across Asia;

R. Pasini, *Landscape Paradigms and Post-urban Spaces*, The Urban Book Series,
https://doi.org/10.1007/978-3-319-77887-7_3

interconnected networks thickening around highly congested metropolitan nodes, while exhibiting secondary urban centers and suburban zones, over geographic portions of North America; finer anthropic structures interconnecting cities, towns, villages and interlocking agricultural territories in Europe. Today over one half of the world's population is considered 'urbanized' (UNESCO 2015, pp. xxi, 7), although the majority spends their life in environments that can hardly be defined 'urban'.[1] An extended literature describes how the sensed border between city and countryside has disappeared (Fig. 3.1).

3.2 The Expanding City

3.2.1 Expanding City Attributes Over Non-urban Space

For a long time now, natural components, such as mountains, valleys, islands, forested plains, interlocked within hyper-expansive anthropic systems and crossed by infrastructural weavings, have increasingly become part of new dwelling scenarios and have grown to a geographic scale. Richard Florida has identified those scenarios from an economist's perspective as novel economic units constituting the "underlying driving forces of the world economy" run by a novel class of creative professionals, and termed them 'mega regions'.[2] His mapping of mega regions on all continents is based on a simple system that interprets light-emission intensity and contiguity data derived from night satellite photographs as a measure of effective socioeconomic integration across a region. While the mere correlation between light emission and socioeconomic-interdependency can misinterpret complex geopolitical and territorial articulations, it is quite efficient at highlighting more elemental dynamics.[3] Alain Thierstein and Agnes Förster's *The Image and the Region* analyzes the 'emerging phenomenon' of 'mega-city regions' and their images with a higher level of disciplinary articulation by mapping more canonic interdependency diagrams of the Munich area (Thierstein and Förster 2008, pp. 14–8, diagrams). The common ground, though, is recognizing the geographic expansion over non-urban areas of basic spatial qualifications traditionally restricted to

[1]"[…] the concept of the urban agglomeration, [referring] to the population contained within […] contiguous territory inhabited […], is favoured over other concepts" (UNESCO 2015, p. 4).

[2]In 1915, Patrick Geddes had described the emergence of regional conurbations that he calls 'city regions' in the chapter 'World Cities and City Regions' of his *Cities in Evolution* (Geddes 1949, pp. 22–31).

[3]With the plan for the creation of the Pearl River Delta administrative unit, north of Hong Kong, englobing 50 million inhabitants stretching from Guangzhou to Shenzhen, including Foshan, Dongguan, Zhongshan, Zhuhai, Jiangmen, Huizhou, and Zhaoqing, the Chinese government has destined financial resources to launch a real-life experiment on the socioeconomic competitiveness of mega-regional entities under a socio-political regime whereby a 'creative class' is not officially contemplated.

Fig. 3.1 Sparrow poacher's trap in an area of *matorral* impacted by infrastructural elements in the expanding outskirts of Monterrey. Flower corollas, berries, and fruits are arranged in a semantic of symmetrical, facial patterns to lure birds into the cage (*photo* rp)

dense urban centers, such as access to social, information, commercial provision or mobility networks. The urban is retreating while the city is expanding.

3.2.2 Geographic Opening

The diffusion of human settlements in very large formations over the last century has evoked an escalating series of denominations: metropolis (monocentric metropolitan environment), megalopolis (multicenter metropolitan environment), and mega city (metropolis on steroids).[4] 'Mega region', the latest term in the urban lexicon, has recently become so prominent as to be widely adopted by the *UN-Habitat's State of the World's Cities 2010/2011* report. In a mega-regional poly-centric network "interlocking economic systems, shared natural resources and ecosystems, and common transportation systems" (borrowing the definition from the *America 2050 Project*), 'non-metropolitan' conditions recur more often than 'metropolitan' conditions, with wild, natural, rural, and suburban areas more extended and even more capacious in absolute population than contained metropolitan centers. As a matter of fact, most of the 200 million inhabitants of the central European mega-regional network[5] do not live in metropolitan environments. The *World Urbanization Prospects 2014* shows how 65% of the urbanized Europeans dwell in cities below 500,000 inhabitants in the face of larger dwelling formations (UNESCO 2015, p. 14, Fig. 10). In the absence of custom data, we can assume that, under European circumstances, the quoted figures are probably coherent with the distribution of 'non-metropolitan' versus 'metropolitan' dwellers. The distinction depends on the level of congestion, speed, intensity, and diversity (Manhattanism), as well as available access to interpersonal exchange, cultural offer, qualified service facilities, as well as supra-local mobility networks of the *cité à la carte* (Sudjic 2007).[6]

[4]'Metropolis', conventionally referring to a city beyond the threshold of 1 million inhabitants, defined by UNO 'middle-size city' up to 5 million; 'megalopolis' (introduced by Gottmann in the '50s for the Great Lakes agglomeration, owing to MacKaye's research of the '20s and '30s and later elaborated by Doxiadis), an urban aggregation with multiple centers of metropolitan level; 'mega city', recent upgraded category for monocentric systems over 10 million, a metropolis on steroids, counting 28 centers worldwide according to UNO's *World Urbanization Prospects 2014*. See also Carbonell's Introduction to *America 2050 Project Report* (Carbonell 2007, p. 5).

[5]The figure includes the population of Greater London (49.1 millions), Greater Paris (14.6 millions), Euro-Lowlands (50.0 millions), Euro-Sunbelt (24.8 millions), Euro Heartland (22.0 millions), Urb-Italy (46.9 millions), as broken down in table 'Megalopolitan City Regions' (Soja and Kanai 2007, p. 63).

[6]See previous Chap. 2 of this book.

3.3 A Novel Urbanity, Its Attributes, and Its Mythology

3.3.1 *Making the Middle Landscape*

Europe is prevalently non-metropolitan as much as America, described by Peter Rowe as "a nation of prevalently suburban dwellers" since at least the '60s (Rowe 1991, pp. 4, 28). While unswervingly convinced of the inherent possibility to re-conduct recent spatial organizations to urbanity, Rowe is thus photographing the prevalence of the territorial expanse characteristic of geographic entities over the concentration of services intrinsic to urban locales. A capital work in the exploration of post-urban environments, *Making a Middle Landscape* describes the manifestation of such a novel environment over the American continent and formulates principles for the making of "a newfound sense of place" (Rowe 1991, p. 264). Rowe sees the 'suburban' as the spatial production generated by the confrontation of 'pastoral perspective' and 'modern technical orientation', which coexist in alienated dualities such as social monolith/social diversity, private commodity/public good, placelessness/place.

3.3.2 *All-American Terrain Vague*

The emergence of a novel American landscape had been illuminated in the groundbreaking work by Christopher Tunnard and his then assistant Boris Pushkarev *Man-Made America: Chaos or Control?* (Tunnard and Pushkarev 1963),[7] and dissected into three functional categories of spatial artifacts, namely suburban housing (referred to as 'low-density housing'), infrastructures for the motorized mobility ('the paved ribbon'), as well as the presence of both industrial and commercial constructions of monumental character in the landscape (respectively 'industry as an element of urban design' and 'commercial facilities in the urban fringe').[8] The groundbreaking novelty of Tunnard and Pushkarev's work is the transversal crossing of these functional axes of housing, mobility, industry, and commerce with an all-pervading urge for the establishment of a novel aesthetics, poised between numbing 'chaos' of cheap heterogeneity and technological 'control' of sophisticated complexity. That novel aesthetics for the vague vastness of

---

[7]According to Pushkarev's own reconstruction appeared in RPA's website in October 2015, the authorship of the work has to be attributed to Tunnard with the exception of Part Three dedicated to the infrastructures for motorized mobility, *The Paved Ribbon*, that he had personally written.

[8]Part Two, *The Dwelling Group*, and Part Three, *The Paved Ribbon*, are dedicated to housing and motorized infrastructures, while Part Four, *The Monuments of Technology*, combines both industrial and commercial uses under the aegis of a novel monumentalization of technology.

contemporary American landscapes, an all-American *terrain vague*, generated by the unrepressed clash of monumental technology and anti-graceful beauty, is what shapes the new space of the 'urban region'.

3.3.3 Pastoral Perspectives and Technological Orientations

The theme of man-made technology and natural beauty clashing to generate modern landscape will also be the major axis bearing the structure of Rowe's middle landscape, under the denominations of 'pastoral perspective' and 'technical orientation' reconciled in the proposed solution of a possible 'modern pastoralism' (Rowe 1991, pp. 216–34).[9] Thus, "modern pastoralism is a symbolic construct" capable to mirror the dynamics of 'pluralist' versus 'majoritarian' instances that, in Rowe's formulation, conform contemporary American society (Rowe 1991, p. 216). The significance of Rowe's elaboration in the broader international debate, and in particular to this investigation, will be treated further.

3.3.4 Mythopoeia

However, Rowe presents 'modern pastoralism' in tight relation with Tunnard and Pushkarev's "poetic doctrine concerned with technological interventions within a rural field", that is "a consistent poetic framework based on careful articulation of man-made elements set against a uniform natural landscape, where material contrasts and spatial intervals play prominent roles". In fact, Rowe regards modern pastoralism as a poetic doctrine or framework governing man's spatial production in the landscape, or in his words "a potentially progressive, critical ideology that can adequately form a mythopoetic context for design in a middle landscape" (Rowe 1991, p. 250).

While Tunnard and Pushkarev's chaos/control dialectics is all internal to the man-made production of space, the unresolved dialectics of 'paradise' and 'pandemonium' (Rowe 1991, pp. 244–8)[10] presented by Rowe in the projected scenario of the modern pastoral is a matter of semantic implantation as much as material production of space and it involves the idea of a possible Arcadian state of nature.

[9]See sub-chapters 'The Pastoral Perspective', 'The Modern Technical Orientation', 'Modern Pastoralism' of Part Three 'Poethics and Making', Chap. 7 'Myths and Masks' in *Making a Middle Landscape*.

[10]See sub-chapter 'Paradise and Pandemonium' in the same book.

3.3.5 The Looks and the Conduct

Rowe resorts to Mumford's words from *The Culture of Cities* to describe the socio-political precincts of the 'middle landscape' as "a collective effort to live a private life" (Rowe 1991, p. 290). The sense of Munford's formulation under-arches Rowe's final invocation for "the establishment of a shared landscape—one that extends well beyond the front or backyard of an individual house" (Rowe 1991, pp. 251–2). Rowe's recipe, meant to requalify the residential, commercial and infrastructural environments of suburbia, focuses primarily on the formal and functional restructuring "of the spatial realms in between [them]". That restructuring is concerned with the embedding of "an appropriate aesthetic form, or 'language', of expression [...] of modern pastoralism" in the suburban mosaic and, by that, realigning the "look of things" and the "conduct of things" (Rowe 1991, pp. 215, 217). Rowe is, thus, proposing a formula analogous to the 'semethic' space envisioned by Almo Farina, produced by both 'semiotic' and 'ethic' configurations (form) and processes (function). This implies the coincidence of linguistic and actional contents, on both the direct level of form and the derivative level of process (Farina 2009).

3.3.6 All-East-Asian Expanses

In later works, Rowe switches to analyzing the contemporary production of space in East Asia, where the emergent expanses of East-Asian cities, large cityscapes that exceed the category of urban space, are referred to as 'architectural territories' (Rowe 2011). The expansion of those cities has been generally concurrent with the creation of "architectural geographies" of spectacular objects, both intended "as tract of land and sphere of action", in as much as their identity is based on a "discourse or accepted manner of engaging in topics" (Rowe 2011, p. 9).

3.3.7 Beijing's Artificial-Lake-Lace

On the one hand, according to Rowe, the spatial production of Beijing responds to "the situational logics and underlying principles of [the city's] persistence of place". The persistence of place consolidates the North–South celestial axis of monuments/values and the East–West axis of modern opening up along Chang'ha (Rowe 2011, p. 17), whose intersection on the Tienanmen Square is surrounded by five circles of ring-roads. Rowe sees Beijing "from the outset as an artifice, for symbolizing a cosmic, social, and moral order, as well as for organizing social and political space with the objective of achieving permanence, harmony and prosperity". Its new compounds, even belittling the old ones, are as large as not to be connected to a recognizable locale, but rather to the composition of a celestial assemblage of values. However, the celestial condition is achieved by means of the "technological temperament", equivalent of the 'technical orientation' that Rowe opposes to the 'pastoral perspective' in the middle landscape. The technological temperament that anciently forged the lace of artificial lakes of Beihai, Zhonghai, and Nanhai, is the same that harnesses contemporary experiments on "the representational possibilities of infrastructure and large-scale entrepreneurial building ventures, […] technical prowess and its display" (Rowe 2011, p. 40).

3.3.8 From Remnin to Lujiazui

On the other hand, Rowe summarizes the recent spatial production of Shanghai with an idea of movement, developed along a non-perspectival (i.e., escaping visual capture), top-down-planned axis. The movement aligns the ancient reterritorialized areas of Yan'an Park and Remnin Precint in Puxi, with the new territories of Lujiazui and Century Avenue, Huamu district and the Century Park in Pudong, on the other bank of the Huanpu River. New Shanghai corresponds in fact to the historically ingrained "aspiration to cross into Pudong" to strike a geographic balance poising over the Huanpu River. Therefore, New Shanghai's *genius loci* is bound to super-local rubrics such as legibility of building layouts, disintegration of architectural objects in spaciousness, and, finally, spatial experience of a scenography (Rowe 2011, p. 74), which refer to a geographic scale.

3.3.9 The Presence of a Mythology

As mentioned in the previous chapter of this book, Rossi's theory of the city openly denies any discontinuity or change 'in quality'—that is in spatial substance—between the urban and the metropolitan conditions. In response to Jean Gottmann's claim for the death of the urban nucleus and its absorption by the economic region (Gottmann 1964), espoused by an entire "school of American ecologists", Rossi asserts that "the reading of the city […] with reference to primary elements, historically constituted urban artifacts, and areas of influence permits a study of the growth of the city in which such changes of scale do not affect the laws of developments" (Rossi 1982, p. 160). The city's deepest substance does not reside in its being separated from the neighboring area by an ancient or modern *urvus*, but rather in its capacity to establish its *locus* (Rossi 1982, pp. 103–7), that is the unique relationship between the site and its buildings. Moreover, the establishment of the *locus* is the coincidence of "the event and the sign that marks it" (Rossi 1982, p. 106), that is the historical sedimentation and its monument—the physical artifact that shapes and preserves a piece of collective memory, being both the place and the myth that describes it. The primary condition for the existence of the city is, thus, not the establishment of a delimitation, but rather the presence of a mythology embedded in its space. Framed in our present circumstances, Rossi's discourse is in favor of the installation of a mythology over the larger scale of the geographic settlement, beyond the urban (*urvus*-based) condition.

3.3.10 Confluence of History and Geography

The *locus* is also the point where history becomes geography, the pulse time of *Chronos* meet the non-pulse time of *Aeon*. In fact, "the history of the city is always inseparable from its geography" (Rossi 1982, p. 95). As we saw earlier, Halbwachs' idea of trans-historical aspirations, developed by a social group dwelling across time on an area, culminates in Rossi's use of the notion of 'expropriations' to indicate the confluence of history and geography. Drawing on Pöete's analysis of historical sedimentation of urban spaces, the idea of the trans-historical aspirations will later develop into Halbwachs' theory of the collective memory bond to urban areas and entail Rossi's osteological theory of the city as fossil memory layered in the *locus*. Incidentally, the idea of the project as reinvention of *locus* does not substantially differ from Marot's idea of the project as *in-situ* mapping of itself.

If Gothic art, according to Henri Focillon, has created France more than geology or Capetian institutions, the matter of a mythology for today's metropolis is that of 'fashioning a new past' and a 'new uniqueness' for 'our existence in the built environment' (Rossi 1982, p. 106).

3.3.11 In Civibus et in Parietibus

Rowe's vast exploration of the emergent territories of East Asia refers the 'production of architecture' to both their 'geographies' and 'discourses'. The production of space is regarded both "as site or parcel of land and as courses of action". Combining flamboyant appearance and conservative approach, the celebratory, spectacular, coarse-grain masterplans of breath-taking scale conjure up the 'narratives of modernization' of expanding populations over geographic regions (Rowe 2011, pp. 198–9). The territorial grounding in action could, in fact, be described as a humongous experiment for the implantation of novel attributes of 'cityness' over geographic regions. The observed phenomenon is the interlaced metamorphoses of two cities, one made of people, the other made of constructions.

Jacques Le Goff comments on the medieval city recalling Saint Augustine's icastic verse "*civitas in civibus est*", according to which the city doesn't stand in its stones but in its citizens (Le Goff 2011, p. 79). Le Goff is describing the early medieval urban imaginary, dominated by verticality both as a mundane power statement and mystic momentum. In terms of Deleuze Guattari's faciality/organization dialectic, Saint Augustine's urbanity could be interpreted as an assemblage of vertical facialities without horizontal structure (i.e., without plan), describing a type of space much closer to a landscape than to an urban plan. Saint Augustine's verse is, moreover, taken from his discourse *De excidio urbis Romae* (Patrologia Latina, XL, 721 ff., sub. 6.6) in which the Punic bishop discusses the losses after the sack of Rome of 410 at the hands of Alaric the king of the Visigoths. The dramatic event that shakes the souls of the subjects of the Western Roman Empire to the point that it is diffusely interpreted as the premonition of the end of the world becomes a symbolic turning-point in the transition between antiquity and the middle ages. From the African coast of the empire, Saint Augustine uses the topic to emphasize the inextricable binary City of God/City of Man, also identifiable with Christianism/Paganism, spirituality/materialism.[11]

3.3.12 Applying Attributes

A new contemporary urbanity can be in sum regarded as the *civitas* spread out, independently from the nominal question of whether it is the city expanding to incorporate the attributes of the 'rure' or the landscape reformulating to incorporate the attributes of the city. The two processes can be envisioned in a serial succession of 'deterritorialization' and 'reterritorialization'. The former process starts when green cities strive to incorporate regenerative attributes of nature and openly question the consolidated regime of contained urbanity. That urban regime had already been destabilized by the expanding boroughs of the industrial city as the mapping of Vienna's growth phases exemplifies.[12] Altogether, this process determines a phase of prevalent deterritorialization. The latter process, instead, the landscape incorporating attributes and 'ethos', or ways, of the city, can be identified with the forces of reterritorialization regaining momentum to reverse instability into a novel regime. As components of the reterritorializing movement, we can regard the action of Manhattanism and landscape urbanism, while sub-urbanism, landscape ecology, landscape aesthetic, and the *scuola territorialista* stand on the deterritorializing front.

[11]Saint Augustine's *The City of God* was composed to confute the diffuse belief attributing the capture of Rome at the hands of the Visigoths to the demise of the ancient pagan cults in favor of Christianism.

[12]In *The Architecture of the City*, Rossi describes Vienna's growth phasing as proposed by Hugo Hassinger (Rossi 1982, p. 66). Rossi also presents Hassinger's 1910 map of Vienna (Rossi 1982, p. 68, Fig. 42).

3.3.13 Capital or Geography

On the one side, Manhattanism replaces nature with the materialization of capital, landscape urbanism replaces the principles of landscape design with the laws of real estate, while, on the other side, sub-urbanism reorganizes the project around the centrality of site's stratification by *in-situ* mapping, and the *scuola territorialista* reacts to the immediate certitudes of globalization tending towards a long-span temporality bordering with eternity. We could say that reterritorializing forces strive to expand urban space over larger extents and therein incorporate some sort of nature, either protected oases or taxidermic prosthesis, while the deterritorializing forces strive to define and incorporate novel attributes of urbanity into the landscape governed by the laws of geography.

Whether by "placing the machine in the garden or the garden around the machine" (Rowe 1991, p. 291), the open question is about reconciling two opposite and simultaneous processes of landscape-making by means of implanting a novel mythology in the geographic space.

3.3.14 About Mythology

According to Rowe, the myth is "a story that expresses or symbolizes deep-seated and exemplary aspects of human existence" (Rowe 1991, p. 244). Mythopoeia, or mythmaking, is finalized to reconciling the inconsistencies between 'abstract principles of social organization and conduct' and 'actual practices' (Rowe 1991, p. 217), or more simply "two otherwise antithetical positions" (Rowe 1991, p. 244). Coherently, the dialectics between the transcendent orientations of romantic pastoralism and laic technocracy permeates the suburban environments of North America, grounding into its geography the confrontation of majoritarian and pluralist democracies that under-arch its post-colonial history. The space of suburbia can be thus imbued with its mythology as a language of legible signs.

3.3.15 Constellations of Sense

By plotting a missing territorial mythology, the semantic layer of Rowe's 'middle landscape' certainly points out opportunities to plot constellations of sense in the mat of North American suburbia, punctuated with infrastructural and service accretions. However, as we noted before in *Life without Monuments*, the 'middle landscape' is also liable to illuminate a much larger field at the global scale (Pasini 2011).

In the middle-realm, "the physical formation, or morphology, of either an urban area or territory" are equalized, indifferently describable "by way of building, urban block, or environmental types" as Rowe notes in his *Methodological Notes* (Rowe et al. 2013). The type is a "unique combination of a suite of characteristics that sets a particular artifact or piece of geography apart from neighboring [space]" (Rowe et al. 2013, p. 127). Not by chance, the type coincides, according to Rowe, with the notion of eco-tope as generally applied in geographic and landscape ecology mapping, and, therefore, assumes a more general and versatile nature.

3.3.16 All Becomes Landscape

Influenced by different forces under disparate geographic coordinates, the super-local middle-realm can absorb the sophisticated territorial fabric in between the traditional European town and its intensively rural land, as well as the novel terrains of Asian mega-cities made of non-visual axes and compounds, the masses of anonymous buildings equally heralding the wasteland of Shanghai's abandoned hinterlands or the fabulous jungles inhabited by monkeys of Hong Kong's New Territories.

In their essay 'Journey Through the Picturesque (a Notebook)', I-ñaki Ábalos and Juan Herreros define 'wasteland' as the "land that has lost its attributes before the approach to the city, that is sterilized as the occupation proceeds". But precisely by virtue of this spoliation of attributes, according to the authors, the wastelands become 'areas of impunity' where the gaze of new social subjects can prefigure new forms of urbanity. This 'new urbanity' means for them the "rediscovery, through architecture, of the contemporary human position in [a] world" where the "opposition between nature and artifice" has dissolved (Ábalos and Herreros 2003, p. 55). The idea of the jungle populated by monkeys refers to an era previous to agricultural domestication and taxonomic ordination of nature, the era of nature's 'inscrutable and threatening force' (Ábalos and Herreros 2003, p. 57).

In fact, both the idea of wasteland and the idea of jungle destabilize architecture's 'anchorage to the place', that is the project as the complete re-transcription of the *locus*, in many possible variations such as the long-hegemonic structuralist or regionalist versions formalized by Rossi and Kenneth Frampton respectively (Ábalos and Herreros 2003, p. 55). From the wasteland/jungle equivalence, so established by the fatal touch of the expanding metropolis, derives the assimilation of the *locus* with the landscape, indifferently natural or artificial to various degrees, which becomes direct 'subject of transformation' through design, no longer backdrop for modernist architectural composition, nor social or cultural palimpsest (Ábalos and Herreros 2003, p. 56). This novel *medium* acquires the attributes of an 'eco-terrain', where architecture's *raison d'être* dissolves and all becomes landscape.

In general terms, the idea of a mega-regional realm is the center of a process of opening up to a geographic dimension.

3.3.17 The Geographic Opened

In fact, Rowe defines the space of the middle landscape starting from the observation of the phenomenon of the expansion of the suburbia over the landscape into the American 'rure' (a neologism derivative of the Latin *rus*, rural space) characterizing especially the 1920s, 1960s, and the 1980s. Rather than to 'suburbs', by 'sub-urbs in rure' Rowe refers to the creation of a 'symbolic landscape', through which transfigured 'mythic themes' of traditional architecture along with the contradictory paradigms of modern 'pluralism in society' have been 'geographically inscribed' in the American space by erasing the distinction between 'urbs' and 'rus' (Rowe 1991, p. 1).

3.3.18 Spatial Operations

The maniphestation of the 'middle landscape' is described as the result of simple spatial operations, namely the "extensive deconcentration of urban population" (Rowe 1991, p. 3) in favor of the prevalently residential expansion of a "uniform network of urbanization moving out over the landscape" (Rowe 1991, p. 28), and the consequent accretion of "agglomerations of [high] intensity and difference in function" (Rowe 1991, pp. 28–9) and the subsequent transformations. The three material operations of 'deconcentration', 'expansion', and 'agglomeration' applied to the American landscape do coincide with the base components of the more general concept of the 'territorial effect' developed between 2004 (Rowe and Pasini 2004) and 2013 (Rowe et al. 2013) with the intention of establishing a general theory of spatial transformation in relation to changes in governing regimes.

3.3.19 Phenomena and Principles

Rowe sees the process of formation of the suburban, fundamentally, as the result of four main concurring phenomena linked to four principles: expansion of residence—the 'ordinary pastoral'[13]; raw mobility infrastructuration—the 'indifferent abuse of land'; accretions of novel polarities—the 'cheap semantic'; and the apparition of the corporate campuses in the wild—the 'techno-Arcadian'.

According to Rowe, the middle landscape appears in four spatial movements. First, the spread of the pastoral residential occurs along railroads first and later vehicular mobility infrastructures. Second, the infrastructures are characterized by bleakly indifferent and functionalist abuse of the territory they cross, while the successive ramification is gated into the semi-private road network of secluded

[13]Quoting Leo Marx.

subdivisions. Third, the establishment of one-stop commercial centers in conveniently accessible positions to serve the multiple necessities of the suburbanites including shopping and leisure occurs at the expenses of the old urban centers. This generates a new species of polarities, where the semiotic of the traditional urban space characterized by permanence converts into the pop overflow of fluctuating ephemerality, mobile and impermanent as linked to instant uses. Fourth, the establishment of corporate headquarters campuses, as modern machines placed in the garden, reclaiming their own introverted patch in the metropolitan mosaic, accrues in spread out areas.

3.3.20 Interstitial Inhospitality

Finally, the interstices in between the patches of the novel geographic mosaic are the space of abandonment (Rowe 1991, p. 245) that Rowe sees as the real opportunity for the regeneration of the novel space and its endowment with the civic attributes of the city.[14]

That interstitial space is liable to produce the modern pastoralism that Rowe advocates for. That interstitial space of relation in between the patches is the significant relational matrix of the middle landscape that Rowe identified as the novelty. That middle landscape is the peculiar trait that does not exist in the overflowing of urban matter off the city walls that has been occurring since antiquity.

3.3.21 Overflows of Different Qualities

Rossi refers to urban matter overflowing and retreating in and out the walls of the Gallo-Roman cities in antiquity, as well as in and out the oval arcades of the amphitheater in Nimes and Arles after the fall of the Roman Empire. The boroughs of Saint-Germain-des-Prés and San Gottardo thrive *extra muros*, on the fields across the Seine from the Petit Pont and outside Milan's Porta Ticinese since the Low Middle Ages respectively (Rossi 1982, pp. 87–8).

However, the urban matter that flows in and out of the boundaries, according to Rossi, remains of the same 'quality'—the same urban substance—as mentioned in the previous chapter. Even in the case of the industrial city of the nineteenth century, when the periphery rarifies, becomes thinner and produces the "*Weichbild*"

[14]Rowe insists on this theme in various points, noting with Leonardo Benevolo the interstitial origin of many revered spaces (Rowe 1991, p. 264) and suggesting the "integration of circulation […] into the sequence of public spaces" (Rowe 1991, p. 286).

of Vienna depicted by Hassinger's 1910 map,[15] he rejects the appearance of an essential gap by which the city and its region would be subjected to distinct laws of space formation. Rossi only admits the manifestation of a distinct tendency towards the city/land divarication during the grave urban decadence of late antiquity. However, the installation of Christian Church's consolidating institutions over the old structure of the imperial territory is the agent that perpetuates the built relics of Greco-Roman antiquity bridging the permanence of its forms into modernity.

To elaborate his osteological theory, Rossi embraces Henri Pirenne's historic analysis of the relationship between the city and its civic institutions. With the reduction of interconnectivity among centers and consequently of commerce (in Rossi's lyric: "exodus of the merchants"), the "great agricultural domains" presided by the castles of feudal nobility exist autonomously from the almost deserted urban centers, whose remaining prestige rests in the "immobile nature of the ecclesiastical office". At this point, in fact, only the installation of the Bishops of the Cristian Church in the ancient centers in ruin preserves the permanence of some forms across the 'anarchy of the tenth and eleventh centuries'.[16] At the "economic and industrial reawakening of Europe", the modern cities of the bourgeoisie can be born again out of the corpses of the old Roman centers that reunite the necessary "conditions of geographic order" at the intersections of the "indestructible roads of Ceasar" (Rossi 1982, pp. 92–3). Again then, the city and its land are one body, as Carlo Cattaneo notes over a century before Heidegger: "The German language uses the same word for the art of building and the art of cultivating: [...] the colonist is a builder (*Bauer*)".

Despite the institutional discontinuity that trades imperial *municipia* for barbaric fortresses out in the countryside and civic homogeneity of the imperial territory for feudal hierarchy, the installation of the Christian dioceses contribute to the preservation of deep forms. The structuration of the Church in the territory allows for the primordial urban semantic to pass over to a next renaissance (Rossi 1982, p. 181, note 4).

3.3.22 *Urban Reforms, Material Continuity, and Semantic Shifts*

While Carlo Cattaneo celebrates 'the region' as man's work of art[17] in the mid nineteenth century, Vienna is demolishing its walls and moats. Constructed after the siege laid by Suleiman the Magnificent in 1529, the walls had resisted the Ottoman

[15]See previous note on Hassinger's Vienna in sub-chapter 'Applying attributes'.

[16]Pirenne powerfully evokes the image in which the Pope makes of Rome the center of Christianity by continuing to live in it when it has been abandoned and, in turn, Rome's historic prestige makes Saint Peter's successor appear larger in his isolation.

[17]See sub-chapter 'Historical continuity of landscape' in Chap. 7 of this book.

expansion in Europe until the definitive Turkish defeat of 1683. Between 1857 and 1872, by order of Franz Joseph I, the city center breaks out and converts the circular glacis into its lofty Ring, a crown delimited by the elegant multiple-tree-lined inner boulevard of the Ringstraße and the functional outer boulevard of the Lastenstraße. Occupied by the new seats of Vienna's institutions, the Ring interfaces the old center with the bourgeois districts consolidating outside the 1529 walls. Barcelona demolishes its walls in 1854 and implements its *ensanche* beyond that perimeter starting in 1860 along the plan by Ildefons Cerdá. In Florence, the demolished urban walls are replaced by circumferential boulevards crowning the old center north and west between 1865 and 1870, inaugurating the thorough urban reform called *risanamento* lead by Giuseppe Poggi.

During the Second Empire, Paris is certainly the model for the metropolitan improvements of the expanding European cities. Between 1853 and 1869, the Prefect of the Seine Department Georges Eugène Haussmann conducts one of the most massive plans of urban renovation recorded in history at the order of Napoleon III. By dismantling the *mur des Fermiers généraux*, a 62-toll-barrier enclosure built on a project by Claude Nicolas Ledoux starting in 1784 and completed right before the French Revolution, in 1860, Haussmann expands the number of the *arrondissements* of Paris almost by the double through the annexation of surrounding Communes. The cyclopean works of sanitation and infrastructuration include the construction of the Gare de Lion and Gare du Nord, Les Halles, the Paris Opera, the metropolitan parks of the Bois de Boulogne west and the Bois de Vicennes east, as well as numerous neighborhood parks over 16 and a half years. The uniformity of façades, materials, and colors precipitates in 80 km of circumferential and axial tree-lined boulevards, producing one of the most characteristic constructs poised between urbanity and metropolitanism: Paris' 'atmospheric perspective'.

According to Rossi, all throughout these urban upheavals, redesigning the major cities of Europe (as for the passage from antiquity to middle age and from middle age to Renaissance), the discontinuity of the institutions corresponds to an inherent continuity of quality and structure, or form, and therefore syntax. Rowe, instead, proposes to map the fluctuation of man's material production of space in coincidence with the fluctuation of the regime governing it. That Rossi's institutions ought to identify with Rowe's regimes *tout-court*, is not granted, but per Rowe, the governing regime of a territory can be read in its language, or semantic, signaling a change in quality and marking an alternative stance. Vienna's *Weichbild*, if not as a fracture in the laws of spatial formation, can be probably regarded as a shift in the quality of the territorial semantic, or mythology, and therefore the forerunning manifestation of the middle landscape.

3.3.23 Territorial Effect

Rowe's *Methodological Notes*, a synthetic compendium of a numerous suite of territorial, urban, and architecture analysis techniques, summarize the concept of 'territorial effect' (Rowe et al. 2013, p. 171), whose elaboration had started with the study on the process of 'territorialization' in Romagna (Rowe and Pasini 2004, pp. 76–82). The idea of the 'territorial effect' is an attempt to capture 'linguistic' characters of space through exclusively numeric factors. By 'linguistic' we refer to a set of characterizations of space that are relational, determined by the relation between and among the parts and the whole, perceptive, depending on variable or multiple observer, phenomenological, proper of manifestations rather than entities, and fluctuant, that is varying across time. Rowe's concise introduction to the methodological notes mentions in fact Martin Heidegger's 'hermeneutic turn', the drop of positivism in favor of relativism, leading to the post-structuralist "spatial turn in geography", where the space becomes "a foreground participant [...] of the agency of action" culminating in Michel Foucault's 'linguistic space' formulation (Rowe et al. 2013, p. 30).

3.3.24 Measures of Linguistic Change

Not by chance the 'territorial effect' is intended to capture the 'significant turning points' (Rowe et al. 2013, p. 171) in the fluctuating regime of 'deterritorializing forces' and simultaneous 'reterritorializing forces' forging our space in a place and a time, referring to the fundamental space-making categories introduced by Deleuze Guattari. The numeric and graphed transcription of phenomenological characteristics is technically performed by registering along the three coordinate axes x, y, z, the measure of three basic classes of spatial transformation, 'expansion', 'evacuation', and 'intensification', occurring in a delimited area in a sequence of discrete periods. The frequency and amplitude of the fluctuation in the sequence of vectors corresponding to each period, show a quantitative expression of the varying 'linguistic' character of space,[18] in analogy with the succession 'geometric plan/topological diagram/permeability graph' (Rowe et al. 2013) after "Michael Ostwald's mathematics of spatial configurations" (Rowe et al. 2013, p. 25).

[18]See in particular chapter 'g. Type and Urban or Territorial Morphology', (Rowe et al. 2013, pp. 127–43).

3.3.25 Numeric But Morphologic

Even though Rowe's methodologies of spatial analysis appear to be colored with a 'positivistic' aura in the array of geo-referenced grid, sieve, dendrogram, and graph techniques for the systematic computational and interpretative processing of data, they are, instead, primarily oriented towards the interpretation of emergent phenomena equally at the scale of geographic, urban, and architectural morphology. Thus, the Euclidian space of a dendrogrammatic matrix, homologous and infinitely extensible like 'kora', is nevertheless loaded with phenomenological, syntactical, or linguistic qualities. Extending over the geographic space to register classes of natural, rural, urban, and architectural spaces, the matrix is, in fact, deployed by Rowe as the applied derivation of the middle landscape construct.

3.3.26 Diverse Mosaics of Uniform Patches

After all, the idea of the middle landscape visualizes a spatial organization of geographic opening. In Rowe, the interpretation of the urbanized network opening to a regional scale resorts to paradigms typical of the geographic or ecological description of a region, which will later become recurrent also in the design field. In *Making a Middle Landscape*, the notion of 'mosaic' is central: the metropolis is "an urbanized landscape made up of enclaves and separate land use fragments" (Rowe 1991, p. 38), where the overall diversity of patches grounds the overall heterogeneity of society and service functions (Rowe 1991, p. 36), while intensifying the physical and social homogeneousness "on the smaller scale of the subdivisions and developments" (Rowe 1991, p. 38). The structure of the 'middle landscape' corresponds to the overall diversity of a landscape mosaic at a large scale, composed of homogeneous eco-topes at a smaller scale. A heterogeneous geographic mosaic of homogeneous 'urban realms' (Rowe 1991, p. 63) transfers much of the complexity of traditional urban morphology, arranged on multiple layers within a contained area, into the unlimited geographic horizontality (Rowe 1991, p. 62).

3.4 The World as One City

3.4.1 Making the World

In *The World According to Architecture: Beyond Cosmopolis*, Hashim Sarkis retraces the origins of the recent interest of designers in geographic aspects focusing on the central concept of 'city-world'. The essay consolidates the notes started in preparation of his course *New Geographies* first taught at Harvard (Sarkis 2011, p. 108, note 1). The debate on the region-forming process at the geographic scale catalyzes now contributions around the *New Geographies* series. Sarkis' essay centers on the concept of 'city-world' defined as the specular antipode to the 'world city' produced by globalization.

As it results clear from the title, Sarkis is referring to the concept of globalized world city, 'cosmopolis', articulated in Edward Soja's monumental book on the 'postmetropolis' (Soja 2000). In the second of his 'six discourses on the postmetropolis', 'cosmopolis' is regarded as the product of the globalization of the city space, what Sarkis calls world city (Soja 2000, pp. 189–232). The broad scope of Soja's 'geohistory of cityspace', the trajectory of urban space in history and geography, starting to learn from Jericho and Çatal Hüyük and ending with exopolitan Los Angeles, exceeds the objectives of this research. But, incidentally, by 'postmetropolis' Soja refers to a 'postmodern metropolis' rather than to a post-urban condition. Like Rowe, Soja makes use of Deleuze-Guattarian deterritorialization-and-reterritorialization dynamics to map the spatial structure as a grounding of the sociopolitical regimes governing a region. However, Rowe's interpretation of the territory is based on the analysis of its physical configuration, backed by graphic and numeric data, focusing on the embedded semiotic text that describes the governing regime. Soja's spatial geography is the result of an anthropological discourse.

According to Sarkis, the city world corresponds instead to the aspiration to "think the world as one architectural entity", or "the capacity to understand and map the living environment" (Sarkis 2011, p. 106).

3.4.2 Radical Lineages

If the world city is proposed as the result of the normalizing modernist model of the International Style, consolidated by the centralizing postmodernist models, the city-world lineage is tracked back to *neo-avant-garde* (or proto-radical) prefigurations. Rooted in Jean Gottman's megalopolis, Constantinos Doxiadis' Ecumenopolis, Yona Friedman's Ville Spatiale, Buckminster Fuller's world-mapping geoscopes, the Situationists' Unitary Urbanism, and Constant Nieuwenhuis' New Babylon, the worldly character of the city world reaches its fully radical unfolding in Superstudio's Supersurface. The city world (as opposed to the world city or cosmopolis or metropolis) recovers "the project of being in the world from the suffocating impositions of globalization" (Sarkis 2011, p. 107) and turns the 'sameness in the world' from 'a sign of poverty of form' to 'an untapped richness' of inspiration. The city world in fact opens into a 'discourse on cosmopolitanism' where 'the subject' is a 'positively nomadic stranger' with world-making powers and where, consequently, the world is 'the scope of individual imagination' (Sarkis 2011, p. 107).

3.4.3 Opposed But Accessible Shores

Sarkis proposes to extend David Abulafia's 'Mediterranean paradigm' to a total reading of the world. In the Mediterranean "geography of opposed but accessible shores", edges consisting of dotted lines of cities and towns are "loosely connected with their [domestic] hinterland" while strongly linked to the opposite, stranger, and often conflictual, shore via trading routes (Sarkis 2011, p. 107). The Mediterranean mercantile centers that generate Abulafia's paradigm of the 'opposed but accessible shores', in fact, are cities and towns such as Cordoba, Valencia, Tunis, Palermo, Mazara, Alexandria, Cairo, forming the Islamic expansion on the southwestern shore, Barcelona, Montpellier, Marseille, Genoa, Pisa, Naples, Amalfi, delineating the Christian territories surrounding the Ligurian and Tyrrhenian Seas, Venice, Corfu, Thessalonica, Athens, Constantinople, Antioch, Cyprus, forming the Byzantine domination over the Adriatic and Aegean areas, as well as Tyre, Sidon, Acre, Jaffa, Jerusalem, the contended strongholds of the Crusaders' states in the Levant (Abulafia 2011).[19]

Abulafia develops the paradigm of the 'opposed but accessible shores' in his monumental book *The Great Sea: A Human History of the Mediterranean* (Oxford 2011). After the shattering in the sixth century of the political and commercial unity of the *mare nostrum* achieved in antiquity by the Roman Empire, "some historians observe decline at the same moments as others detect expansion" in a 'painfully slow' process of re-integration of the Mediterranean as early as 9th and tenth century (Abulafia 2011, p. 241). Even acknowledging the 'resilience' of the 'Byzantine East' and the basic continuity of the Islamic lands stretching from Syria to Portugal, the Mediterranean ranges "enormous regional variation" well beyond the 'puzzle' of the Christian West (Abulafia 2011, p. 241). What Abulafia describes as the 'Third Mediterranean' could be regarded as the process of formation of the urban Mediterranean, precipitating in particular between the tenth and the fourteenth centuries.[20]

---

[19]See maps on pp. 260–1, 272–3, 288–9, 320–1, 336–7.

[20]See in particular Part Three, Chap. 2 'Crossing the Boundaries between Christendom and Islam, 900–1050' (Abulafia 2011, pp. 258–70). There, Abulafia reconstructs the weft of 'a Mediterranean society', using Shlomo Dov Goitein's definition derived from the exploration of the Cairo Genizah collection of Jewish traders' documents. As Abulafia explains, the Ben Ezra synagogue of Old Cairo was rebuilt in the eleventh century by the Jewish population of 'Palestinian' liturgy (ancestor of the liturgy used by Italian and German communities and rival to the 'Babylonian' liturgy adopted by the Sephardi), incorporating a storeroom, *genizah*, on an upper floor only accessible through a ladder. The *genizah* was used to stuff discarded documents, mainly commercial papers, bearing Hebrew characters whose destruction would have been sinful. The Cairo Genizah collection of traders' documents is a chaotic assemblage that, after Goitein, Abulafia calls 'the opposite of an archive'. However, the Cairo Genizah collection replots that mosaic of cross-boundary connections conforming the varied mosaic of Mediterranean society coincident with its geographic space. The analogy between the geographic vision derived from the apparent chaos of the Cairo Genizah collection and the explorations of the geographic space through apparently illegible 'big data' is quite manifest.

3.4.4 *Smooth Passages*

In Part Three, Chap. 5, Abulafia reconstructs the voyages across political boundaries in the Mediterranean, undertaken in the second half of the twelfth century by Jewish rabbi Benjamin of Tudela and Muslim high-ranking bureaucrat Muhammad ibn Ahmad ibn Jubair,[21] respectively from Navarre to Jerusalem and from Granada to Mecca.[22] Benjamin courses the Christian lands of eastern Iberia down to Barcelona, reaches the lands of the Franks at Marseilles and from there by boat arrives in Genoa. In spite of some discrepancies between the account and the route provided by Abulafia,[23] Benjamin probably proceeds from there overland, skipping navigation to Pisa due to its constant state of conflict with the Genoese, visiting Rome on his way to Norman Bari. Through Corfu, he then leaves Christendom to start his march, overland again, across the Byzantine dominion towards Constantinople. Over Cyprus, again he crosses the crusaders kingdom in the Levant, leaving vivid descriptions of Acre, before reaching the holy city of Jerusalem. He probably fancies a further trip to explore Mesopotamia, whose descriptions might be product of fantasy. Southwards from there, crossing into the Muslim sphere, he reaches Alexandria. From Alexandria, he can easily embark for his return journey home via Sicily. Noting down eager commentaries about all the places and peoples encountered, the Jewish traveler Benjamin offhandedly passes from Christendom to Byzantine Empire, from Constantinople to the Christian Levant, to just smoothly cross over to Islam and then sail back to the Christian West again, over the waves of the Mediterranean.

3.4.5 *The Formation of the Urban Paradigm*

The voyage of ibn Jubair is more troubled due to a series of accidents culminating in a shipwreck off the port of Messina during the wintry return lag. However, despite the tribulations of the journey, his crossings over the Christian–Muslim divide occur with equal ease. Ibn Jubair's description of the 'civil and respectful' search on the Muslim travelers at the Christian customs at Acre is even contrasted with the 'harsh and unfair' treatment received by them from the Muslim officers in Alexandria. In spite of a conventional diffidence among ethnic groups, and beyond episodes of friendly relationships between individuals, all his accounts report of unabridged commercial convergence of the ethnic groups of the Mediterranean mosaic, essentially Jews, Christians, and Muslims, in the teeming ports that are in

---

[21]In literature more commonly transliterated as Ibn Jubayr.

[22]See the classic editions of *The Travels of Ibn Jubayr, being the Chronicles of a Mediaeval Spanish Moor Concerning His Journey to the Egypt of Saladin, the Holy Cities of Arabia* (Broadhurst 1952) and *Viajes de Benjamin de Tudela* (De Llubera 1918).

[23]See map (Abulafia 2011, pp. 306–7) and Cfr. *Viajes de Benjamin de Tudela*.

the process of regenerating the spatial identity all around the Great Sea. The description of crowds of diverse origins and their intense commercial interchanges in Barcelona are paralleled to analogous circumstances encountered in Constantinople, Acre, and Alexandria (Abulafia 2011, pp. 304–17).[24]

What is described by Abulafia here, is, thus, a process of formation of a geographic region indissolubly rooted deep in the urban paradigm, despite different regimes, religions, and magnitudes, that so drastically impacts the way in which man sees the surrounding world still today.

3.4.6 Geographic Aesthetics

Sarkis recognizes in the modes of imprinting of the Mediterranean as an urban region, a geographic modality (landscape urbanism would term it 'change in ethos') of space production that can be opposed to that of globalization. That is a modality of exploring the geographic dimension alternative to the globalizing expansion of the world city over the geographic dimension. In fact, while modern geography, the discipline risen from the "history of discovery and colonization" with the aspiration to be the synthesis between "the physical, the economic, and the social" has disintegrated into other disciplines, according to Sarkis, today, "a geographic aesthetic dominates formal pursuits" and designers are called to "transform larger contexts" addressing "infrastructures, urban systems, rural and regional questions" (Sarkis 2011, p. 108, note 1).

The special interest in the natural realm is witnessed by Philip Ursprung's *Natural Histories* (Ursprung 2006) characteristic of our time is well expressed by Josep Lluís Mateo: "whereas in the recent past the paradigm by which architecture was measured was the city, now, the collective reference surrounding our design activity is the relation to nature". "After Rossi and Koolhaas, a manifestation of the operational impracticability of nostalgia and delirium, the city appears as a second nature"—Mateo glosses (Mateo and Sauter 2014, p. 8).

3.4.7 Vice Versa

The Soja/Sarkis issue can be interpreted more as of a dialectics rather than an exclusive alternative. Despite nominal juxtapositions, no sharp separation stands between the idea of the city that becomes global, Soja's world city, and the idea of the world that become one city, Sarkis' city world, inasmuch as the processes that generate the latter imply symmetrical processes that generate the former and *vice versa*. Also, although the phenomena described by Soja and Sarkis are specific to

[24]Chapter 5 of Abulafia's book is titled 'Ways across the Sea, 1160–1185'.

the contemporary upheavals investing global space, their generative principles are more general. The core of the question is the idea of a globalized cityscape, cosmopolis, that contains a city world beyond cosmopolis and *vice versa*. The dynamics of deterritorialization (city world) and reterritorialization (world city) are never exclusive, but always reciprocal and simultaneous, only characterized only by the prevalence of one or the other force.[25] To this respect, it might be useful to introduce an idea of cosmopolis, construed from a disciplinary perspective separate from that of design.

3.4.8 *Ancient Cosmopolis*

> "The spectacles produced in the vast arena [of the Colosseum] paraded the city's mastery of the world." The diversity of the spectators in the Colosseum, as the diversity of the city's population, "embodied the vastness [...] of the Roman empire [and] Rome's power to draw people to itself over distances almost unimaginable, from cultures thrillingly alien." (Edwards and Woolf 2003, p. 1)

In *Rome the Cosmopolis*, historians Catharine Edwards and Greg Woolf set off to investigate "the nature of the relationship between the city and the world" in Roman classical antiquity via a composite fresco of essays (Edwards and Woolf 2003, p. 3). The idea of 'Cosmopolis' is not necessarily aligned with the conventional concept of cosmopolitanism that permeated the ancient Roman empire as it permeates our globalized reality. Exceeding the implications of cosmopolitanism, Edwards and Woolf's work excavates a set of timeless nuclei from Roman antiquity, pertinent to the idea of the city world as a contemporary continuum in need of a mythology.

In fact, according to Edwards and Woolf, the identification between Rome and the world goes beyond the idea of the 'cosmopolite', the "wise man, transcending local attachments to identify with all of humanity as a 'citizen of the world'". They use Ovid's *Fasti* 2.684 stating that "the world and the city of Rome occupy the same space" to start a deeper exploration of the relationship Rome/world summarized in the critical term 'Cosmopolis', unknown in ancient literature (Edwards and Woolf 2003, p. 3).

3.4.9 *Synecdoches and Metonymies*

A constant paradox is at the basis of 'Cosmopolis'. 'Cosmopolis' rises from the irreconcilable binary of the City simultaneously dominating and epitomizing the

[25]"Deterritorialization is always double [exhibiting] a deterritorializing force and a deterritorialized force" with the relative roles of "expression" and "content". We can identify the mentioned 'deterritorialized force' with a 'recessive force of reterritorialization'. See 'Theorem Five to Theorem Seven' (Deleuze and Guattari 1987, pp. 306–7).

world. Rome is synecdoche for the world, being its dominating head, *caput mundi*, but also metonymy for the world, as the City tends to absorb its totality. While the whole of the world is to be found in Rome, Rome drains off the world of all its beautiful possessions and talented inhabitants. Ancient Rome is a *cité à la carte*, where all the produce and products of every region of the world are available.

'Cosmopolis' is a mixed system of coexisting dynamics of violence and absorption. While Rome is extending its rule over the world (domination) and spoiling its treasures (depredation), the City is simultaneously refining its civility and culture (adaptation) and expanding the attributes of its urbanity (diffusion, redistribution), primarily its *ius* and security.[26] The domination/adaptation dialectics marks its climax with the ostension of Agrippa's *imago mundi*[27] in the northern Campus Martius, displaying in and before the city the entirety of the conquered world, in correlation to the *forma urbis*,[28] the cadastral map of the city carved on marble panels to be exposed in the *forum pacis*.

Edwards and Woolfs conclude that the global flourishing of 'Cosmopolis' would also mean the eclipsing of the world, in line with Strabo's commentary in *Geography* 5.3.8 according to which, after seeing the monuments of the City, "you would easily become oblivious to everything else outside. Such is Rome".

3.4.10 Layer and Relayer Space with Identitarian and Heteronomous Myth

While converting into 'Cosmopolis', Rome is fulfilling its destiny born of spontaneous and forceful miscegenation of Italic, Asian, and Etruscan peoples, casting the 'heterogeneous Rome' as the urban other to the 'autochthonous Athens'. To stand against change, the City needs deeper roots in antiquity, but, at once, to "persist as a faithful epitome of the world" (Edwards and Woolf 2003, p. 9), the City has to keep pace with the multiple changes across it.

In order to become 'a fixed point in the Cosmos', the City 'layers and relayers' its space with identitarian and heteronomous 'myth and history', to produce 'a theater of memory' and at the same time 'a stage for the actions of the future' (Edwards and Woolf 2003, p. 8).

In her essay 'Incorporating the Alien: the art of conquest', Edwards leads us through a colorful voyage across the statuary of ancient Rome, as numerous and

[26]Edwards and Woolfs quote Aelius Aristides praising in the second century "the Romans for extending the security associated to urban life through the empire" and Rutilius Namantianus considering in the fifth century that, by expanding justice over the lands, the Romans "have made a city of what was once a world" (Edwards and Woolf 2003, p. 3).

[27]Dating back to the last years of the first-century BC, often cited by Pliny the Elder in his *Natural History*, and probably the base for the Peutinger Table as for many other Roman and medieval maps (Friedman and Figg 2000, p. 8).

[28]Beginning of the third century.

multiform as to compose a 'second population' of the City rivaling with the human population (Edwards 2003, pp. 44–70). By that, Edwards draws an alternative topography of 'Cosmopolis', centered on the process through which "alien was appropriated and incorporated into the fabric of the City, until it became a permanent mnemonic of the empire" (Edwards and Woolf 2003, p. 19).

Roman statues are intermixed with statues looted from the conquered territories of the empire forming a physiognomic and cultural babel. As for the obelisk of Heliopolis, depredated with the spoils of Augustus' victory at Actium to be re-erected in the Campus Martium to commemorate the conquest of Egypt, urban elements from the ruled world undergo a process of 'appropriation' and 'recontextualization'. It is a process of urban miscegenation through which "the city had absorbed the world" (Edwards and Woolf 2003, p. 2), poised between counteracting aspirations to diversity and Romanness, exoticism and identity, alien and self.

3.5 Chapter's Conclusions: Mnemonics of the World

The idea of 'Cosmopolis' as the 'mnemonic of the world' is a poetic transfiguration of a possible contemporary space, compatible with both Soja's dialectics of globalization through the fractal terrain of postmetropolis, as well as Sarkis' autopoiesis of a geographic space of complexity across the city world. Not by chance, Pirenne's idea of the coincidence of 'municipal life' and 'national life' as well as 'city' and 'Imperial system' in Roman antiquity, is the base on which Rossi founds his notion of the permanent continuity of the spatial formation, city and its land, city and its region.

References

Ábalos I, Herreros J (2003) Journey through the picturesque (a notebook). In: Mostafavi En M, Najle C (eds) Landscape urbanism: a manual for the machinic landscape. Architectural Association, London

Abulafia D (2011) The great sea: a human history of the mediterranean. Oxford University Press, Oxford

Broadhurst RJC (1952) The travels of Ibn Jubayr, being the chronicles of a mediaeval Spanish Moor concerning his journey to the Egypt of Saladin, the Holy Cities of Arabia, London

Carbonell A (2007) Introduction. En America 2050 project report. Lincoln Institute of Land Policy and American Plan Association, New York

De Llubera IG (1918) Viajes de Benjamin de Tudela. Madrid

Deleuze G, Guattari F (1987) A thousand plateaus: capitalism and scihzophrenia (trans: Massumi, B). The University of Minnesota Press, Minneapolis [1980]

Edwards C (2003) Incorporating the Alien: the art of conquest. In: Edwards C, Woolf G (eds) Rome the cosmopolis. Cambridge University Press, Cambridge

Edwards C, Woolf G (eds) (2003) Cosmopolis: Rome as world city. In: Rome the Cosmopolis. Cambridge University Press, Cambridge

Farina A (2009) Ecology, cognition and landscape. Springer, New York [2006]
Friedman JB, Figg KM (eds) (2000) Trade, travel, and exploration in the middle ages: an encyclopedia. Routledge, UK
Geddes P (1949) Cities in evolution. Williams and Norgate, London [1915]
Gottmann J (1964) De la ville d'aujourd'hui à la ville de demain. La transition vers la ville nouvelle. Prospective (11)
Le Goff J (2011) La Città Medievale. Giunti Editore, Firenze
Mateo JL, Sauter F (2014) Earth, wind, air, fire. Actar/ETH, Barcelona, Zurich
Pasini R (2011) Life without monuments. Forum A+P (11), 78–93
Rossi A (1982) The architecture of the city. The MIT Press, Cambridge ([Italian 1966] ed.)
Rowe P (1991) Making a middle landscape. The MIT Press, Cambridge
Rowe P (2011) Emergent architectural territories in East Asian cities. Birkhäuser, Barcelona, Basel
Rowe P, Pasini R (2004) Territorialization in areas of Romagna, Italy. Harvard GSD, Cambridge
Rowe P et al (2013) Methodological notes on the spatial analysis of urban formation. Harvard GSD, Cambridge
Sarkis H (2011) The world according to architecture: beyond cosmopolis. In: Jazairy EH (ed) New geographies n. 4: scales of the earth. Harvard GSD, Cambridge
Saint Augustine (1841–1865) De excidio urbis Romae. In Migne JP (ed), Patrologia Latina, XL, 721 ff., sub. 6.6 [after 410]
Soja E (2000) Postmetropolis: critical studies of cities and regions. Wiley, Hobocken
Soja E, Kanai M (2007) The urbanization of the world. In: Burdett ER, Sudjic D (eds) The endless city. The Urban Age Project by the London School of Economics and the Deutsche Bank's Alfred Herrhausen Society. Phaidon, London
Sudjic D (2007) Theory, policy and practice. In: Ricky Burdett EDS (ed) The endless city. The Urban Age Project by the London School of Economics and the Deutsche Bank's Alfred Herrhausen Society. Phaidon, London
Thierstein A, Förster A (2008) The image and the region. Making mega-cities regions visible. Lars Müller, Zurich
Tunnard C, Pushkarev B (1963) Man-made America: chaos or control?. Yale University Press, New Haven
UNESCO (2015) World urbanization prospects 2014. United Nations
Ursprung P (ed) (2006) Natural histories. Lars Müller, Zurich

Chapter 4
The Symbiotic Field in Ten Behaviors

Abstract This chapter identifies ten behaviors, or operative methods, adopted by contemporary designers confronted with problems of spatial reform in regions of the contemporary continuum. The analysis proceeds through the compared anatomy of a set of case studies, tackling spatial problems in a diversity of scales and contexts. The cases include a majority of implemented projects, but also speculative visions, figurative compositions, and performances to explore the tangible and intangible layers of the contemporary continuum. By arraying operative approaches spanning from metabolist to geophilosophical, the ten behaviors conjure up a 'symbiotic field' layering the ecological, societal, psychic, aesthetic, cultural, semiotic, geographic, and mythological: built macro-ecologies of geographic scale; artificial micro-ecologies implanted into the metropolitan platform; ecological installations raising awareness through aesthetic performances; cybernetic scenarios preconizing the control of natural/anthropic metabolism; atmospheric assemblages dealing with the cultural and the transient; experiments of psychic and sensorial manipulation of the landscape; explorations of the transculturation process in a utilitarian modernity versus symbiotic culture dialectic; formation of alternative spaces in the folds of the language; geopolitical plans representing the geography of power; geophilosophical excavations aiming at reimplanting a mythology into the landscape. Furthermore, the chapter analyzes the radical utopias that constitute the recurrent reference for many of today's visions of the contemporary continuum. The text also argues on how the utopian visions of the 60s and 70s have unawares become today's quotidian living contexts.

Keywords Macro-ecologies · Micro-ecologies · Cybernetic control
Atmospheric assemblages · Sensorial manipulation · Utilitarian modernity
Geopolitical · Geophilosophical · Agronica · Tirana

R. Pasini, *Landscape Paradigms and Post-urban Spaces*, The Urban Book Series,
https://doi.org/10.1007/978-3-319-77887-7_4

4.1 Introduction

Despite sharing a perspicuous interest in the great osmosis of theory contents between the field of design and numerous disciplines spanning ecology, geography, sociology, statistics, economics, philosophy, aesthetics, logistics, infrastructural engineering and territorial management, this research also intends to ground its notions in concrete reality (Fig. 4.1).

This chapter analyzes an anthology of design cases loosely fitting categories compatible with the landscape construct described in Chap. 7 'Reconciling Cultural and Cognitive' and previous theoretic elaboration. By avoiding any tendency to rigid prescriptiveness, this selection constitutes a brief compared anatomy. The selected projects decode the interaction of natural and artificial, physical and intangible systems from individual standpoints.[1]

Borrowing an image from Sébastien Marot, each project represents an original trajectory to venture out in the landscape continuum "like someone who plunges into an unknown jungle with no map at hand" (Marot 2003, p. ii). Moving from disparate starting points, the assemblage of the different trajectories, altogether, conjures up the missing map.

A time of transition from metropolitan to regional, from urban to post-urban, from anthropic/natural to symbiotic, that is a 'crisis' in its original meaning of transformation, offers the opportunity to form radical questions and the liberty to interpret them: Can we come to an armistice with our planet and abandon architectural construction as we have practiced it? Can we stop creating individual buildings locally integrated with their sites and suppress the concept of artificial prosthesis implanted onto the territory? Can we, on the contrary, coordinate our efforts in the collective pursuit of a general Fullerian 'pattern of integrity' for identifiable territorial organisms? Can we stop designing buildings and start engineering functioning environmental entities such as a valley, a mountain, a forested plain, and a watershed, including their exposed and hidden systems, capable of supporting life: flora, fauna, and humans?

4.2 Ten Behaviors Before the Contemporary Continuum

4.2.1 Geotechnics or Applied Geography

Kongjian Yu's large-scale landscapes realized in China with Turenscape deal, in various conditions and contexts, with the theme of hydrogeological systems

[1]Several authors of the designs analyzed further on, participated in the TeleTalks conference series curated at the University of Monterrey in the period 2014–16, some events in collaboration with Laura Cipriani of IUAV. Extensive documentation of the conferences is collected in a separate publication.

Fig. 4.1 View of 'weak-metropolitan' territory looking southeast from the cafeteria's terrace of Polis University at km 5 on the Tirana-Dürress autostrada, exhibiting an assemblage of infrastructures, large industrial plants, cultural institutions, fine residential fabric, agricultural fields, cattle (*photo* rp)

Fig. 4.2 Qunli Stormwater Park, Harbin, Heilongjiang Province, China, Turenscape 2011 (*photo* courtesy of Turenscape)

redesign. These landscapes are real implementations of 'geotechnics as applied geography', according to the synthetic definition by Benton MacKaye. MacKaye in fact attributes its conception to Patrick Geddes, who reportedly coined it during a joint hike in 1923.[2] Applied geography, however, or geotechnics, corresponds to the 'machinic' quality applied to nature (Fig. 4.2).

These cases of 'applied geographies' are a symmetrical 'analogous' to the English picturesque, which "began to dream of identifying itself with nature" (Marot 2003, p. 20).[3] On the one hand, the picturesque recomposes the surface of a phenomenological scene from an imaginary landscape of geotechnical stability. On the other hand, geotechnics redesigns, from underneath, a functioning environmental machine that is capable of autonomously covering itself with a phenomenological skin. And that skin takes the form of a possible landscape.

While the artificial topography at Stourehead is implemented to freeze in situ a combination of scenes from Claude Lorraine's Mediterranean visions, the artificial hydrogeology at the Houtan Park in Shanghai is engineered to kindle a process of transformation on a multilayered piece of landscape, which is let free to evolve into diverse phenomenological phases.

4.2.2 Machinic Micro-Ecologies

Ecosistema Urbano designs machinic apparatuses capable to reproduce fundamental functions of natural systems, creating microenvironments with controlled

[2]"Geography [...] is a descriptive science; it tells what is. Geotechnics is applied science; it shows what ought to be" (MacKaye 1950–1, p. 439). In the same text, Benton MacKaye also accounts of how he had retrieved in a Webster's International Dictionary of the '40s the famous definition often erroneously attributed to him: "Geotechnics—the applied science of making earth more habitable". That definition reportedly orchestrates in the most effective form a series of three terms that had been agitating in his mind for about 40 years: 'geotechnics', 'habitable globe', 'greater habitability'.

[3]Marot is quoting John Dixon Hunt.

conditions. Two projects focus on the machinic reproduction of vegetation and its microclimate-regulating functions aimed at rehabilitating peripheries of relatively high density with insufficient urban installations. The designers implemented Air Trees focusing on the control of climate conditions in a strategic site in order to promote social interaction among alienated communities. Variations of the original type were also developed by elaborating on the potential of ludic interplay to reconstitute the social fabric among youth, or focusing on the potential of an interlaced information platform connecting local and global levels. The installation of Air Trees is strategically planned to support the gradual growth of an abundant provision of natural vegetation. These machinic micro-ecologies are not envisioned as an alternative to nature, but rather as means to further the reappearance of living systems within derelict man-made precincts and eventually to bridge the nature/city divide (Fig. 4.3).

4.2.3 *Aesthetic Prostheses for an Operative Metabolism*

Studio Roosegaade's Smog Free Tower has the appearance of a technological monolith, whose slowly retracting doors offer cunning glimpses into an inner technology veiled in obscurity. It is presented as a hippie totem installed amid a neglected open space invaded by returning vegetation and surrounded on all sides by the skyline of some indistinct urban or metropolitan formation. The tower is a large air-purifying apparatus supposed to work outdoor at the scale of the metropolis. Its filter captures the carbon dioxide in the air and its hidden machinery compacts it into small solid cubes, which, like 'diamonds', can be later be mounted onto jewelry pieces. The totem catalyzes around its mystic presence liberated social

Fig. 4.3 MediaTree, Madrid Ecoboulevard, social occupation of the space, Ecosistema Urbano, 2007 (*photo* Emilio Doiztua, courtesy of Ecosistema Urbano)

Fig. 4.4 Smog Free Tower, Daan Roosegaarde Studio, 2015 (*photo* courtesy of Studio Roosegaarde, CC)

dynamics reminiscent of those orchestrated in Superstudio's imageries of the Supersurface series (Fig. 4.4).

Conceptually, the Smog Free Tower is a prosthesis installed onto the hybrid metabolism of a natural/artificial system, groundbreaking for its elemental straightforwardness. However, its practical applicability appears to be evidently improbable. Just like for other projects by the same office, the Smog Free Tower's most manifest aims largely remain those of an aesthetic performance that moves consciences and triggers good practices.

4.2.4 *Cybernetic Utopias*

Part of Project Solana, the Montenegrin contribution to the Venice Biennial 2016 curated by Bart Lootsma and Katharina Weinberger, ecoLogicStudio's Open Aviary envisions a technological network capable to control the complex ecosystem of the abandoned saltworks of Ulcinj, embracing soil, water, flora, fauna, and humans. A computing station instantly processes the data stream from ESA's Sentinella2 satellite, forming a high-resolution mapping of the geographic area around the saltworks. An algorithm is capable to separate the data stream into various information levels, rendering instant images of the changing environmental conditions at the site, spanning from mineral, to hydraulic, biochemical, and

Fig. 4.5 Solana Open Aviary bio-digital substratum, ecoLogicStudio, 2016 (*image* courtesy of ecoLogicStudio)

vegetational processes. Additional data are fed from ancillary systems capable, through sensors and detectors, to monitor the movements of the aviary fauna at a local, regional, continental, and intercontinental scale. The station has the potential to implement spatial transformations through a set of controlled robots on the field in response to various environmental conditions. Similar stations around the world can be interlaced in a global network (Fig. 4.5).

If brought to the ultimate consequences, the project casts the promise of a future pacification of the environment, when the constant satellite monitoring of infinite variables returns the total photograph of the planetary metabolism to a wise computing center, capable to regulate the overall course of things within the anthroposphere. A cybernetic utopia seems to be implied, where, by virtue of responsive systems, the control of immaterial information bestows the control over physical reality.

4.2.5 *Atmospheric Assemblages*

A tiny, very poetic, gif movie illustrates the implemented ExpoGate by Alessandro Scandurra in front of the esplanade of the Castello Sforzesco in the heart of Milan

Fig. 4.6 ExpoGate, Universal Exposition Milan, Alessandro Scandurra 2014 (*image* composition of still frames from gif file courtesy of SSA)

(Fig. 4.6). The ExpoGate is at once a physical place and a place of otherness (Foucault 1984). The pavilion is a physical architectural apparatus installed on today's dense relational platform of the castle, once the interface between the urban and nonurban conditions is historically delimited by the lost city walls.

Housing the info-point of the Universal Exposition of Milan 2015, the ExpoGate is also by its function a wormhole raised to the square. It opens, in fact, a virtual passage into the dislocated space of the Expo fairgrounds, which is, in turn, another physical place as well as another 'otherness', mentally transferring the visitors into disparate locations around the world, associated with the pavilions.

The landscaping multiplicity and profundity of this simple design is revealed by the tiny motion file recording the clumsy gait of a dog, comforting urban fauna, and crossing the frame of the image. Its gracious trajectory precipitates the layers of city, construction, atmosphere, air, rain, and fountain spray until they intersect, merge, and coexist in a phenomenological and relational construct, the atmosphere, which is the diaphanous material of Scandurra's work.

The metal structure of the ExpoGate was recently dismantled. Not gently, but swept away with bulldozers and trashed. This does not diminish the pavilion's aura, but rather rarefies its atmospheric qualities. Its ephemerality is elevated to that of the mist floating over the fountain.

4.2.6 The Psychic Sphere

François Roche and new-territories.com's cyber-fantasies envision blurred entities whose nature stretches from object to ecosystem. Roche has directed his research towards the limits where the categories of natural and artificial hybridize until they become indistinct. He works on informational, perceptive, and psychic layers to conjure up manifestations of 'new territories', which coincides with the name of the collective. The explicit identification of subject and object, whose relational play produces the landscape, not only expands over the geographic extent of physical space but also replots the physical point by point onto the virtual territory of the worldwide web represented in the '.com' suffix in the name of the office.

Mythomanias is a stream of consciousness, interweaving lyric texts, generative narratives as well as posthumous commentaries, with sequences of video-recorded projects, installations, and experiments (Lab M4 et al. 2015). The doors of perception

Fig. 4.7 Alterate State, dissemination of a cocktail of physiological pheromonal substances, New-Territories, 2013 (*photo* courtesy of New-Territories)

are opened, the psychic state of the beholder is altered, and epiphanies of anomalous subjects/characters/users traverse the wilderness of the Thai jungle or the wasteland of Bangkok's abandoned skyscrapers with improbable trajectories, converting the physical/psychic/virtual environment into a multiple landscaping construct (Fig. 4.7).

4.2.7 Creole Horizons

Arctic Adaptations: *Nunavut at 15* by Lola Sheppard and Mason White's Lateral Office can be described by referring to the linguistic paradigm of creole.[4] Presented in the Canadian Pavilion at Rem Koolhaas' Venice Biennial of Architecture, *Arctic Adaptations* marks the 15th anniversary of the founding of Canada's newest territory, Nunavut. The 'creole' is the language of the new natives, generations after the more or less violent fusion of ethnic groups following substantial relocations. Not a language one can adopt by will, it is a cultural confluence in which one is

[4]For 'creole' and 'miscegenation' see subchapters 'La idea del mestizaje' and 'El nacimiento del criollismo' (Bernal 2015, p. 46, 312). See also next Chap. 5 in this book.

born. The work in fact accepts the exhibition director's request to address the theme of *Absorbing Modernity: 1914–2014* and to show "the process of the erasure of national characteristics in favor of the almost universal adoption of a single modern language". *Arctic Adaptations* records the rapid confrontation of Canada's newest territory, Nunavut, with modernity, termed "a transition 'from igloos to Internet' in 40 years" (Sheppard and White 2017). Revealing powerful native traits of adaptation and resilience, the project records the capacity of native culture to elaborate modernity in a novel creole. The Arctic creole is traced in the mapping of reformulated architectural artifacts and novel intersubjective dynamics, as well as in the redesigned cloud of tools and practices surrounding man in the arctic icescape (Fig. 4.8).

4.2.8 Semiotic Landscapes

Perry Kulper uses the graphic transcription of the process of design to explore it as a stratified accumulation of information. Aggregations of heterogeneous entities, afferent to disparate categories ranging from physical presence to intersubjective perceptions or individual memories, find a virtual space where surprising conceptual, combinatory, and logic relations can interweave. It is the exclusively human 'eco-field' associated to the cultural trait, described by Farina (2009).

The trans-categorical profundity of the composition expands the potential of the design process to the investigation of a state of reality as well as to the prefiguration of surreal alternatives. It is an investigation of the virtual and relational space of language, which interferes with reality in the phenomenological and intellectual framing of the physical space, revealing additional unfolding. Kulper's earlier bidimensional explorations, such as *David's Island* (Fig. 4.9), have expanded into tridimensional experiments, such as the flying paint experiments realized with Nat Chard (Kulper and Chard 2013). By mixing virtual and physical elements in visionary compositions, their landscaping character results enhanced. These experiments fathom 'the folds of difference', as Kulper puts it, that is, landscapes of

Fig. 4.8 Kimmirut, Canada, Arctic Adaptations: Nunavut at 15, Lateral Office 2014 (*photo* Billy Aakavak, courtesy of Lateral Office)

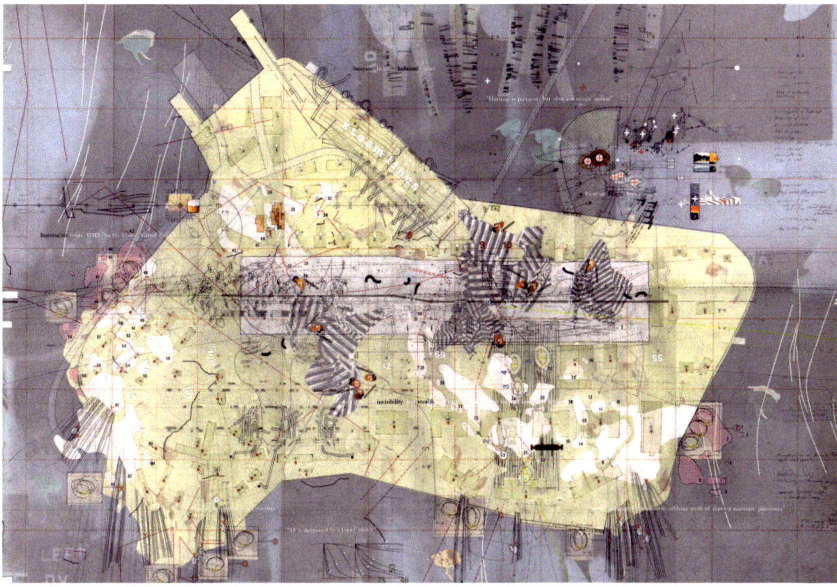

Fig. 4.9 David's Island Competition, strategic plot, Perry Kulper, 1996–7 (*image* courtesy of Perry Kulper)

instrumentality, made up of diverse entities ranging from graphic, informational and conceptual, to physical, natural, or artificial, all transcribed into shared platforms of communication.

4.2.9 Geopolitical

By synthesizing the spatial vision of *Albania 2030 Manifesto* (Aliaj et al. 2014), Besnik Aliaj deploys a geopolitical reading of the present and future of the Albanian territory throughout a series of sharp diagrams drafted by Eranda Janku. This manifesto aimed at prefiguring a broader scenario of territorial governance for the country also stimulates a reflection on the compatibility of geopolitical aspirations with the prerogatives of an architectural territorial vision. The expanding and simultaneously thinning role of the designer and planner, everyday less involved in construction and more concerned in addressing participative processes of socio-spatial requalification, shifts towards an intentional political agency. The geopolitical field is often one where politics invites architectural prefigurations as vehicles for its cannier maneuvers. The manifesto revolves around the idea of territorial development, de facto identified with that of economic development, based on the definition and enhancement of a hardcore horizontal hierarchy, able to

Fig. 4.10 Rural area in the municipality of Belsh, Albania, from Albania 2030 Manifesto, Aliaj et al. (2014) (*photo* courtesy of Eranda Janku)

enweave natural, social, and cultural heritage at different scales. The resulting landscape is a geopolitical scenario in which soft components are prevalent (Fig. 4.10).

4.2.10 Geophilosophical Excavations

The Espai Barberí in Olot is the in situ mapping of a geophilosophical field, poised between stratification and construction. Many of RCR's designs are dominated by the concept of 'excavation'. By virtue of excavating, the project is embedded into the geologic stratification, which is the repository of long-term history beyond the historical time of man. From that superhuman succession, the project's monolithic metal slabs surface from blast furnaces as from the cave of Vulcan.

Springing out of the ultraslow-forming geology, the project cuts through the geophilosophical accumulation of the landscape. That landscape is man's history, layered and relayered over topography to convert into geography (Fig. 4.11).

Fig. 4.11 View from vegetated patio into Espai Barberí, RCR Arquitectes' self-designed office, Olot, Spain (*Image* courtesy of Hisao Suzuki)

The excavation of the geological is intersected with the excavation of the osteological. The fragmented skeletons of architectural artifacts interspersed in the geological material are rediscovered as the remnants of remote existences in an archeological repository. Moldy brick walls, semi-crumbled septa, rusty cast iron columns and trusses, with their naïf decorations of a nostalgic past, are redeemed from the condemnation of waned ages and recuperated to an extended life that seems miraculous, next to crystal panes and heavy bronze plates. The landscape and the ruins of man are interlocked with the obscure foliage of the woods in a benign labyrinth enjoying endlessness. The woods are admitted into the inhabitable space, or even constitute the very stuff with which the project is built. It is the construction of the Jüngerian woods, *den Wald zu bauen*, the place of hermitage of the Anarch.

The project is firmly set at the fault between orography and the plains, where the dark woods that cover the Pyrenean hillsides with the slow succession of a dense vegetation meet the open fields of fat crops and fine grapes. The Espai Barberí, made of woods and ruins, is an incessant excavation, observation, recuperation, and reinvention of a human/natural mythology in the landscape of sparsely inhabited regions.

4.2.11 A Bundle of Feebler or Bolder Routes: The Symbiotic Field in Ten Behaviors

On an axis stretching from metabolist to geophilosophical, these projects line up in a quite comprehensive sequence of the operative approaches towards contemporary landscape's manipulation. The symbiotic field that is being conjured up altogether is a composite of ten distinctive behaviors, or methods, intersecting the ecological, societal, psychic, aesthetic, cultural, semiotic, geographic, and mythological: (1) macro-ecologies at the geographic scale are obsessively focused on the issue of built ecologies and productive landscapes (Qunli Stormwater Park); (2) artificial micro-ecologies implanted into the metropolitan platform are about ushering the return of pacified nature as much as about reviving social interaction (Madrid Ecoboulevard Air Trees); (3) ecological installations aim at raising awareness about natura/anthropic metabolic processes, while staging prevalently aesthetic performances (Smog Free Tower); (4) staging of cybernetic scenarios aims at envisioning the control-through-information of natural/anthropic metabolism in regions of the anthroposphere (Solana Open Aviary); (5) atmospheric assemblages deal with cultural as well as meteorological mists, the exploration of space through the exploration of the transient time of the meteors[5] (Milan ExpoGate); (6) sessions of collective—human–human and human–machine—psychic alteration test the sensorial manipulation of the landscape (Alterate State); (7) anthropological explorations of the transculturation process following the colonization of a symbiotic culture by a utilitarian modernity aim at reactivating social dynamic capable of novel identitarian dynamics (Nunavut at 15); (8) proliferating linguistic environments aim at reformulating alternative spaces in the folds of the discourse (David's Island); (9) geopolitical plans deals with the geography of power (Albania 2030); (10) excavations of the landscape as a geophilosophical accumulation, relayering geology, vegetation, ruins, osteology aim at the reimplantation (recovering and reinvention) of a mythology (Espai Barberí).

These designs explore the present transitional phase in the discourse on human settlement on the planet in search of a new aesthetic paradigm, framing the cultural perspective of inhabitable space, as well as an implementable model at a time when the dismantling of the urban-centered vision is advanced.

Altogether, their trajectories trace a bundle of feebler or bolder routes in the novel geography of the expanded field of landscape and conjure up a multi-perspectival collective observation. The overall result is meant to fathom fragments of our relationship to nature and project new scenarios of symbiosis, combining aggregates of artificial and living systems, psychological and physical spaces, and machinic and cultural landscapes.

[5]About the 'time of the meteors', positioned by Michel Tournier between the extreme 'speed of the asters' and the extreme 'slowness of geology', see Chap. 6 in this book.

4.3 Past Utopia and Present Everyday

4.3.1 Failed Leaps into Artificial Paradises

The contemporary behaviors before the landscape that we have discussed share common foundations in the imageries propagated by the radical collectives in the 60s and 70s.

Supersurface (a video, installation, and design piece by Superstudio), in analogy with other 'non-work continuum' (Tafuri 1972, p. 398) precedents, such as Constant Nieuwenhuys' New Babylon and Archizoom's No-Stop City, reveals a utopian scenario of technological advancement whereby the primordial natural platform becomes obsolete, totally replaced by a man-made grid able to support forms of collective dwelling. These days, Superstudio's Supersurface is enjoying a particularly favorable revival, being celebrated worldwide in numerous publications, exhibitions, and even recreations of the original installation at the famous MOMA exhibition of 1972. Radical architectural prophecies of a future field reconciling nature, technology, and humans were instantly torn down as a "private leap into the sublimated universe of artificial paradises" by Manfredo Tafuri from within the very catalog of the exhibition where they were showing, for delusively resting on a utopian technological advance and the consequent emancipation from work (Tafuri 1972, p. 394).[6]

Andrea Branzi's Agronica (Branzi et al. 1995), or Weak Metropolis, from 1995 marks both the last evolution of that kin and the beginning of the dissolution of the myth of metropolitanism. An anthropic infrastructure expanding over vast extensions, thinning, and dispersing its artificial limbs of factories, Agronica interweaves with agricultural patterns to form a double-productive dispersed territory. Despite its ideological content, disciplinary abstraction, and allegations of ecological unsustainability, Agronica's productive territory is, according to Branzi, sustained by an alternative conception of ecology, whereby transcendent intersubjective values redirect the ordinary physical sustainability balance.

4.3.2 Tirana: A Case of Quotidian Agronica

Agronica is less a utopian vision than one might think. The mixture of rural and industrial activity characterizes large portions of the European province. In the case of the Albanian territory around Tirana, the brutality of the clash is made more stunning by the archaic characters of its rurality congealed by the communist

[6]Generally referring to radical architecture, or neo-avant-garde, or counter design, or anti-utopias, Tafuri chiefly addresses his repeated invective against Achizoom's No-Stop City and secondarily Superstudio's Continuous Monument.

regime *vis-à-vis* the rough impatience for a process of privately-fancied-of Balkanic modernization.

In the last decades, since the transition to democracy, following the long immobility administered over the territory by the communist regime, Tirana has undergone a fierce process of both uncontrolled densification and expansion. The interstitial spaces of the consolidated urban organism have been chaotically occupied while the rural and agricultural environments outside the urban boundaries have been extensively littered with the informal proliferation of incongruous constructions. The heritage of urban and rural systems resulting from century-long processes of territorialization was abruptly threatened of almost complete erasure by the emerging phenomena related to either the aware or unconscious adoption of 'metropolitanist' models. Hegemonic 'metropolitanist' models have conveyed a universal promise of flamboyant happiness along with a delusory offer of congested modernity. More recent paradigms, born from the analysis of the global economic and productive system, have drawn attention from the role of the metropolitan centers to that of the much broader mega-regional formations encompassing natural, rural, urban, and sprawling features, much in accord with models of low-density anthropic/natural coexistence.

Expanded areas around Tirana are quite representative of a contemporary Agronica condition. The zone stretching from the Paskuqan Lake plains up to Mount Dajt, encompasses parts of the city proper, agglomeration of various natures, from suburban to informal, the breathtaking system of the natural water bodies, the surviving relics of the traditional rural fabric and the natural theater of the mountains. The hybrid landscape on the Tirana-Dürress Autostrada presents analogous features. The infrastructural networks interwoven with the rural fabric and industrial structures of varying size, mixed with agricultural and cattle-breeding activities, configure the 'double-productive' landscape preconized by Branzi. The garish marquees of commercial brands pop up amidst higher education facilities, factories, grazing fields, haystacks, and cows, offering a vision into what could be defined unresolved 'modern pastoralism' (Rowe 1991, pp. 215, 217).

4.4 Chapter's Conclusions: Domestic Landscape in Biotechnological Field

However, if we accept Branzi's hint to a mystic form of sustainability, the present-day uncontrolled plotting of an interface between production and information, and the subsequent ongoing dematerialization of the production system, might offer an opportunity for a 'new domestic landscape' (Ambasz 1972) within a wider biotechnological field of symbiotic character.

Advanced biotechnological applications for the creation and regulation of a symbiotic landscape made of man-made and natural systems interfaced at the genetic level, as well as the level of information/command exchange is a breathtaking open abyss that future generations will have a chance not only to explore but even fill up, invent, and engineer. Cybernetic landscape scenarios of hybrid systems exhibit today just a shy shadow of their future potential.

References

Aliaj B, Janku E, Sotir D, Allkja L (2014) Albania 2030 Manifesto: a national spatial development vision! Polis University, Tirana

Ambasz E (ed) (1972) Italy: the new domestic landscape. MOMA, New York

Bernal R (2015) Mestizaje y criollismo en la literature de la Nueva España del siglo XVI. Fondo de Cultura Económica, Ciudad de México [1994]

Branzi A et al (1995) Symbiotic metropolis Agronica. In: Susani M, Manzini E (eds) The solid side. V+K Publishing, Rotterdam, pp 100–121

Farina A (2009) Ecology, cognition and landscape. Linking natural and social systems. Springer, New York [2006]

Foucault M (1984) Of other spaces: Utopias and Heterotopias (trans: Miskowiec J). Architecture/Mouvement/Continuité [1967]

Jünger E (2006) Il filosofo e l'Anarca. Intervista a Ernst Jünger. In: Volpi F, Gnoli A (ed) L'ultimo sciamano. Bompiani, Milano

Kulper P, Chard N (2013) Pamphlet architecture 34: phatoming the unphatomable. Princeton Architectural Press, New York

Lab M4, Lacadée C, Roche F (eds) (2015) Mythomanias: a series of psycho-architecture case studies. Punctum Books

MacKaye B (1950–1) Growth of a new science, geography to geotechnics (series of articles). The Survey (magazine)

Marot S (2003) Sub-urbanism and the art of memory. Architectural Association, London

Rowe P (1991) Making a middle landscape. The MIT Press, Cambridge

Sheppard L, White M (2017) Many norths spatial practice in a polar territory. Barcelona, New York, Actar

Tafuri M (1972) Design and technological Utopias. In: Ambasz E (ed) Italy: the new domestic landscape. MOMA, New York

Part II
Dealing with the Composite

Chapter 5
Miscegenation: Culture- and Region-Forming Processes

Abstract While contained urban space, as reviewed in the first section of the book, suggests series of binaries deriving from the original dichotomy city/nature, a geographic expansion favors the commingling of composite or opposite principles. The second section of the book analyzes ways in which elective or imposed coexistence of diverse agents on a geographic field kindle a process that evolves from juxtaposition to reciprocal permeation and eventually miscegenation, both in culture- and region-forming. This chapter describes apparently unrelated phenomena as concurrent manifestations of miscegenation: (1) the auto-reformulation of Josephine Baker's self-agency, unfolding in the sequential transformations of her celebrated banana skirt resonating in artistic expressions of the early-twentieth-century primitivist modernism; (2) the formation of open space ensembles in the colonial cities of Nueva España from the fusion of the Mesoamerican and Mediterranean ideas of place-making; (3) the implementation of the monumental metropolitan boulevard of Reforma in Mexico City, linking the colonial historical center to the pre-Hispanic hill of Chapultepec and its venerated forest. These productions of urban and metropolitan beauty stand in sheer contrast with the violent transformations of the territorial land patterns of the hinterlands of Mexico, displaced derivatives of the land expropriations of the French revolutionary period that shape the dreary face of the new nation through the transition from *colonia* to *estado nacional*. The text draws parallels between the cultural and spatial implications of the colonizing action brought forth by conquering over conquered groups in a territory.

Keywords Miscegenation · Colonization · Culture-forming · Primitivism Region-forming · Expropriations · Reforma

R. Pasini, *Landscape Paradigms and Post-urban Spaces*, The Urban Book Series,
https://doi.org/10.1007/978-3-319-77887-7_5

5.1 Introduction: Geography, Discovery, Colonization, and Miscegenation

In *Mestizaje y criollismo en la literatura de la Nueva España del siglo XVI*, Rafael Bernal identifies the origins of miscegenated *entendement* in the awe of the Spanish explorers-conquerors before "la presencia de América" (Bernal 2015a, p. 323),[1] with its unknown natural entities of unprecedented magnitude and variety (Bernal 2015a, p. 73)[2] to put names on. The natural, botanic, zoological, and ethnological observations exact new terms since the very beginning, imposing the expeditioners to draw on the tongues of the (otherwise erased) cultures of the Caribbean islands as well as on Nahuatl and Maya parlances on the mainland.[3] It can be argued that the urge to put a name on a new thing triggers a process of reciprocal contamination that culminates in the first utterance. When the name is spoken, a miraculous aura is spelt over a location, where physical and mental worlds make contact, astounding the subject.

The aghast contemplation of the 'presence of America' is, thus, followed by mental interpretation, hazardous exploration, physical interaction. Spatial exploration is intertwined with cultural and anthropological exploration. According to Bernal, when Hernán Cortés sets sail from Cuba towards the mainland in 1519, not only is he already intentioned to convert the short commercial expedition he had been commanded into sedentary expansion, but he is also aware that the great undertaking of the colonization of the New World is conditioned to a necessary miscegenation between Spaniards and indigenous. He will phrase that condition in the principle of the 'conservación de los naturales' (Bernal 2015a, p. 114),[4] which implies coexistence and reciprocal safeguarding between two belligerent ethnic groups.

According to Bernal, Cortés had been predisposed to cultural miscegenation from his witnessing of the tragic fate of previous Spanish foundations on the mainland, devastated by unknown illnesses and surrounding hostility of unknown natives. More deeply in his soul, the century-long conflict/exchange with the Moors, losing and reconquering lands in a permanent region-forming succession, separating and connecting the two opposite shores of the Mediterranean, had already eradicated racial scruples from the Spaniards (Bernal 2015a, p. 56).

Flocks of missionaries, humanists, and soldiers, then, could pour over the new lands with the purpose of colonizing as well as understanding. Columbus' motive for the exploration was that of tracing new commercial routes and possible bases for trades, while Cortés imagined the territorial expansion of the nation. The drive for the appropriation of lands, resources, and riches of various sorts was inseparable from the momentum pushing towards the ethnological and anthropological

[1]In English: 'the presence of America'. For the Spaniards, it was still the 'Indias', but Bernal refers to the sparked trajectory that will turn the continent into the New World.

[2]See entire sub-chapter 'Las cartas de Colón y el asombro de las Indias'.

[3]See entire chapter 'Las voces caribes de los Españoles'.

[4]In English: 'preservation of the natives'.

investigation. The conquest of the New World was, thus, generating modern geography, made up of economy, space, and anthropology. The driving countenance of the Mediterranean model of the 'opposed but accessible shores' had been transplanted into new continents.

The anthropological succession described by Bernal lines up Indian Spaniard, *criollo*, that is creole, and *mestizo*, that is 'miscegenated'. At first, the displaced Spanish colon transformed by the experience of the New World lives in constant dissatisfaction for purportedly insufficient recognition of the pioneering work done to the benefit of the nation. Then, the offspring from Spanish parents born on American land is imprinted with a sense of superiority towards the natives and a diametrical complex of inferiority towards the Europeans, which feeds anti-indigenist resentments. The final achievement of a dispassionate miscegenated awareness, which sees its own identity before the world in the integral assimilation of indigenous, Spanish, and creole characters, is identified by Bernal in the Mexican society of the nineteenth and twentieth centuries (Bernal 2015a, p. 323).

This latter euphemistic representation probably corresponds to a rhetoric compatible with Bernal's diplomatic office in the service of Mexico on various shores of the Pacific and to his reactionary, corporate political positions.[5] The sociological reality is of course else, in a nation still segregated on the basis of economic circumstances as well as somatic characters. However, the model proposed by Bernal to interpret the linguistic and cultural succession during the process of Hispanization of America is surprisingly compatible with the scope of contemporary design practice intended as a contribution to the region-forming process at a geographic scale.

In particular, Hashim Sarkis' reformulation of David Abulafia's Mediterranean paradigm of the 'opposed but accessible shores' constitutes a powerful tool in the interpretation of the evolving landscape.[6] It represents a form of embedment of a collective and anonymous design agency in the region-forming process, way more convincing than any proposed version of mild-determinist metabolism.

Several are the correspondences between the structures and contents of Abulafia's *The Great Sea* (Abulafia 2011) and Bernal's *El Gran Océano* (Bernal 2012), which for the amplitude of the matters exceed this research and offer great room for future work.

---

[5]Bernal's political activism in the filo-fascist, nationalist, and rhetorical *movimiento sinarquista* had lead him to sojourns in the Mexican prisons around 1948. His monumental work *El Gran Océano* was mainly elaborated in the years 1960–65, when deployed to the diplomatic missions in Honduras, the Philippines, and Peru. Through his missions, he had the opportunity to explore firsthand *'el campo histórico inteligible'* of the transculturation processes, between the opposite shores of the Pacific rather than across their aquatic medium. His cultural action in the diplomatic service focused on the promotion of international collaborative studies, resulting in a considerable 'transpacific intellectual flow' between Mexico, Peru, the Philippines, China and Japan. (De Maria and Campos Castelló 2015) As a counterpoint to his diplomatic and political stances, in his contemporary fictional production, Bernal is otherwise a sharp castigator of his homeland's hypocritical rhetoric (Bernal 1969, 2015b).

[6]See Chap. 3 of this book.

5.2 Cultural Miscegenation

5.2.1 A Glittering String of Artificial Bananas

Since her Parisian years, the glittering figure of Josephine Baker has been tied to the world of architects through subterranean exchanges, recurrent and unexpressed, which despite intense critical exploration have retained an essential obscurity. In her essay 'A House for Josephine Baker', the Australian feminist architecture theorist Karen Burns argues that 'criticism's nominal project' (Burns 1997, p. 70), that of deciphering spaces for 'unimpeded visibility', would fail in interpreting the architectural constructs generated in Baker's name and their exterior/interior dialectics, if attention is not turned to a "discourse [on] the biography of Josephine Baker" (Burns 1997, p. 63). We should, in fact, go beyond Burn's suggestion to align architecture and Baker's life, and align architecture and Baker's very body (Fig. 5.1).

Born in Saint Louis, USA, from a couple of vagabond music performers, she is an abused live-in domestic at 8, a street child scavenging food from the garbage at 13, a street corner dancer at 14, and a last-chorus-line dancer[7] in Harlem at 15, where her clowning dance moves[8] gain attention. At age 19, offered an adventurous contract by an emissary of the Theatre des Champs-Élysées, Baker sets sail to Paris along with a company of 20 African-American revue musicians and performers. *'La Revue Nègre'* is an all-black musical show, whose conception is generally attributed to Fernand Léger's excitement before the African sculpture exhibition at the *Exposition des Arts Décoratifs* of Paris 1925 (Sweeney 2004, p. 44). In the wake of the *'negrophilie'* (Negro fever) fascinating the Parisian Avant-Garde culture of the '20s, the modernist elite ends up attending rehearsals and premiere in large number.[9]

Perfectly fitting the openly erotic canons of the Parisian music halls of the '20s that indulge colonial curiosity focused on exposed nudity, bizarreness, exoticness and sensuality of black feminine (but also masculine) bodies, Baker happens to be unexpectedly promoted from her chorus girl post to sharing the role of *vedette* few whirlwind days before the premiere (Roueff 2006, p. 67).

There she opens the show doing splits in the air while being carried on comedian-dancer Joe Alex's shoulders, with nothing but a pink flamingo feather

---

[7]The comedy chorus girl at the end of the line is a traditional persona of black vaudeville. She cannot get the step upon entering the stage, but gets it better than anyone else 'breaking the place out' in the *encore*, as narrated by Patrick O'Connor in the BBC documental *Josephine Baker* (Broughton and Phillips 2006, frame 5′40″).

[8]Interviewed in the same documental, Brenda Dixon Gottschild enumerates the 'shimmy', 'mooch', 'mess-around', and 'Charleston'.

[9]Sweeney mentions Man Ray, who took photographs of the cast, Jean Cocteau, Robert Desnos, Blaise Cendrars, Francis Picabia, Paul Guillaume, Fernand Léger, Kees Van Dongen (Sweeney 2004, p. 45).

Fig. 5.1 Josephine Baker in *Un Vent de Folie* wearing a glittering banana string, Folies Bergères 1927 (*photo* Lucien Waléry, PD legal notice at end of chapter)

Fig. 5.2 Photo of Baker and Alex performing the 'Danse de Sauvage' number of the Revue Nègre showing at the Théâtre des Champs-Élysées, probably taken during the promotion of the show before the press in 1925. The iconic power of the image relies on the embodiment of the unconscious fancies of the primitivist cultural milieu. Alex' figure, the combination of brute force and supreme agility of an ideal natural state, effortlessly supports the flying evolutions of Baker's plumed nudity while drawing the aerial fresco of nature's invincible beauty (*photo* courtesy of akg-images)

between her legs for the canonic *'loufoque'* (crazy) choreography.[10] Her body shaking in a tiny kilt of plumes, climbing onto Alex's ebony mass and swinging around it, closes the show with the other expected *topos*, the erotic choreography, interpreted in *'la danse sauvage'* (Fig. 5.2). Perpetuated by Paul Colin's posters and, later, his lithographic album *Tumulte Noir,*[11] *'la danse sauvage'* projects the entire company in a European tour until an abrupt interruption in Berlin. There, enticed by the director of the *Folies Bergère* Paul Derval, Baker breaks her contract and returns to Paris to feature as the uncontested diva of the production *La Folie du Jour*, where she first appears dancing in a banana skirt as Fatou in 1926. Fatou is a gracious sexy little savage that descends from a tree in the jungle to tease a white explorer by shaking her bananas (Barnwell 1997, p. 85). That first banana skirt with

[10]"She made her entry entirely nude except for a pink flamingo feather between her limbs; she was being carried upside down and doing the splits on the shoulder of a black giant. Mid stage he paused, and with his long fingers holding her basket-wise around the waist, swung her in a slow cartwheel to the stage floor, where she stood… She was an unforgettable female ebony statue." (Flanner 1925)

[11]On Paul Colin's *Tumulte Noir* album of 1927, see (Dalton and Gates 1998).

Fig. 5.3 Baker in a Fatou-model rubber banana skirt, during her Copenhagen tour of 1928 (*photo* Tage Christensen, courtesy of Ritzau Scanpix)

the primitivist looks of a natural tuft of small bananas jiggling curved towards the inside in various rows (Fig. 5.3), will be replace in next year's production *Un Vent de Folie*. A sophisticate string of longer bright yellow bananas, elegantly spaced, pointing outwards, and beautifully lined with sparkling crystals, comes to embody the manmade improvement of the natural, the work of art elaborating on the primordial, modern primitivism. A Freudian representation of the playful empowerment over a multitude of erected phalli, the banana costume will become a *topos* of

Baker's artistic self-agency in numerous variations (Sowinska 2005–06).[12] The artful string of beautiful bananas, however, can be regarded as a striking manifestation of the tension towards a *mètissage* (miscegenation), equally permeating the Parisian *avant-garde* art scene and the European late-colonial society that had expanded its violence over vast exotic territories and was now returning a gaze imbued with a different desire.

5.2.2 Modernist Primitivism

According to Oliver Roueff, Baker's seduction is based on three unresolved ambiguities elaborating upon eroticism, namely savage and civilized, feminine and androgynous, sensuality and innocence (Roueff 2006, p. 70). The black feminism and cultural studies theorist Mae Henderson also speculates that the vast interest instantly raised by Baker's primitivist performances are to be situated in the context of the French *belle-époque* colonialism colliding with the explosion of modernity at the beginning of the XX century.

5.2.3 Colonial Geographies

The colonial tradition of 'ethnographic display' of primitive manifestations precipitates, in fact, into the modern 'erotic/parodic performance' of the *Revue Nègre* producing a 'powerful iconography' of the *métissage* of 'modernist primitivism' (Henderson 2003). Henderson, thus, identifies the binary voyeurism/exhibitionism with the binary colonizer/colonized, as the obsession of the former for looking (male) opposed to the desire of the latter to be looked at (female).

On a larger geographical scale, the three ambiguities suggested by Roueff can be equally referred to as binaries plotting the colonial dialectic between colonized and colonizer that generates the miscegenated space of new continents. Those binaries, in fact, plot the geographic divergence between the 'rhizomatic' and the 'arborescent' models, the East and the West envisioned by Deleuze and Guattari (1987, p. 18). The various articulations of that divergence, such as 'Oceania' versus

---

[12]Sowinska describes Baker's banana skirt as a collective creation to which one of Paul Colin's posters for the Revue Nègre must have contributed with a fundamental intuition. There first appears a transfiguration of Backer's body in animalistic countenance, veiled by a tuft of bananas. Jean Cocteau's artistic direction and many other characters peopling the music hall scene of the '20s, confusedly mentioned in Baker's autobiographies including herself, must have variously contributed to the gradual evolution of the costume from the jingling rubber tuft to the glittering bananas string.

'Europe', the 'steppe-and-garden' or 'desert-and-oasis' versus the 'forest-and-field', the 'replanting-of-offshoots' versus the 'sow-and-reap', correspond to the fracture through which liberated sexuality gets rid of the genealogic imperative (Deleuze and Guattari 1987, p. 18). The process of 'becoming-' through masochist practice indicates sexuality as the field of deterritorialization/reterritorialization parallel to geographic imprinting.

5.2.4 *Mythic Persona*

The string of artificial bananas precipitates, then, the transfiguration of Baker, stricken by worldwide celebrity, into a mythic persona and a symbol of transcultural miscegenation. The fact that 2 years later, after long vocal training, she is ready to play the lead role in Jacques Offenbach's operetta *La Creole*, sanctions the formidable transformation from '*petit danseuse savage*' to '*grande diva magnifique*',[13] or the 'creole goddess'.

5.2.5 *Cet obscure objet du désir Colonial*

Baker was the object of desire of an entire generation, the muse of among others Hemingway, Scott Fitzgerald, Picasso, Dior. Carole Sweeney elaborates on the relation between 'aesthetic modernism' and 'imperial capitalism' to note that the "cosmopolitan access to subordinate cultures that European modernism enjoyed [was simultaneous to] the concentration of capital in colonial territories" (Sweeney 2004, p. 48). According to Sweeney, 'radical innovators of architectural form' such as Le Corbusier and Adolf Loos among the modernist *avant-garde*, were inspired by Baker's 'denaturalized' body for its participation in both modernist and primitivist aesthetic (Sweeney 2004, p. 48).

Cultural historian Brenda Dixon Gottschild anchors the body of Baker to a spatiality that rhizomatically reconnects the microscale of the object-subject to the macro-scale of geography by stating that "[Baker's] black dancing body [disrupted the picture of] the monarchical, hierarchical, vertically aligned, body of European ballet". Baker's 'denaturalization' is performed by overriding the verticality of the sedentary subject with the obliqueness of the nomadic subject through movements that go back to the primordial salvage and the baboon that swings on a liana. While overthrowing the European mode, the 'awkward becomes beautiful', the 'silly savvy', and 'everything sexy' (Broughton and Phillips 2006, frame 12′30″).

[13]A popular quote from British singer Shirley Bassey.

By Deleuze-Guattarian terminology, to move from the sphere of art to the geophilosophical plateau, Baker's 'denaturalized' body can be better said to be the 'body without organs', or the 'line of flight', that rhizomizes and miscegenates the territory of modernity.

5.2.6 The Colonial Perspective of Landscape Urbanism

Our view over the landscape is tightly related to colonial perspectives. Mostafavi and Najle's 'manual for the machinic landscape' (Mostafavi and Najle 2003), the collection of essays and projects intended to be foundational to the movement of landscape urbanism, was reviewed earlier as a regressive counterpart to Rowe's concept of the 'middle landscape' and Sarkis' idea of the 'region-forming' process, while formulating a contemporary idea of landscape *continuum* in this research.

In it, Hight's 'Portraying the Urban Landscape: Landscape in Architectural Criticism and Theory, 1960–Present' (Hight 2003), far from being the announced historical-critical reconstruction, represents the systematic manifesto of the movement following Corner's conceptual foundation. Hight's analysis of landscape relies on the concept of 'orientalism' derived from Edward Said. "The 'other' [...] plays a constituent part in constructing the identity of the imperial power" (Hight 2003, p. 25, note 4), in order to describe the relationship between landscape practice and architecture. The parallel between the binaries 'colonizer/colonized' and 'architecture/landscape practice' results extremely pertinent to our matters.

Our earlier discourse on 'miscegenation'[14] constitutes a construct specular to that of the 'orientalism', that is the possible role-reversal in the architecture-landscape contamination of ethos advocated for by landscape urbanists. In fact, if landscape is the 'adjacent other' to architecture, then, in line with the 'orientalist' formulation, it is architecture, the colonizer, the term that is being redefined by the 'adjacent other', the landscape, the colonized. The actor of the self-reformulation is in this case architecture, the colonizer.

5.2.7 Miscegenation

On the other hand, the 'miscegenation' model implies the specular contamination of the colonized, landscape, from exposition to the colonizer, architecture. In the 'miscegenation' model, the colonized is the protagonist. The colonized is, in fact, the agent of its own reformulation on the basis of the hybrid knowledge acquired.

Rather than exclusive alternatives, the 'orientalism/miscegenation' dialectic is more of a reciprocity, recording prevalence of movement in one direction or the

[14]See: sub-chapter '6.1 Geography, discovery, colonization, and miscegenation' in this book.

other. After all, the colonizer/colonized dynamics can be seen in Deleuze-Guattarian terms, where both deterritorialization and reterritorialization processes imply the secondary presence of the other. But, despite this reciprocity, the construct proposed in this research is, thus, that of ‘miscegenation’. It can be described as a self-operated reterritorialization of the colonized, following its deterritorialization produced by the exposition to the colonizer. That is a prevalence of emancipatory cosmopolitanism of the colonized over auto-absolutory exoticism of the colonizer.

We therefore advocate for the prevalence of the virtues of ‘miscegenation’ over the hypocrisy of ‘orientalism’, that is, in strictly disciplinary terms, the prevalent self-redefinition of landscape practice over its colonization by transplants of architectural ‘ethos’.

5.2.8 *Self-reformulations*

As of Baker, we could conclude that she is self-reformulating herself after the deterritorialization operated on her by colonialism. As the colonized, Baker is the main subject of the unrequested disruption operated on her by colonialism (heteronomous deterritorialization), as much as she is the main subject of the self-reformulation (autonomous deterritorialization) that she is operating on herself by redefining her characters. We can consider that ‘miscegenation’ records the prevalence of self-deterritorialization of the colonized, for which heteronomous disruption is the triggering factor. Both disruption (heteronomous deterritorialization) and self-reformulation (autonomous deterritorialization) combine in a process of deterritorialization culminating in the self-reformulation of Baker’s persona and body.

5.2.9 *Unsolicited Attentions*

In 1928, Loos designed for Baker a famous unbuilt black-and-white-striated house on the Parisian Avenue Bugeaud, for which neither trace of a commission, nor of reaction exists. Le Corbusier fell to Baker’s feet the next year, in 1929, during his transatlantic journey from Buenos Aires to Bordeaux on ocean liner Lutétia. A photo shows him in a black-and-white-striated costume staring at Baker during a fancy-dress reception. He would later insistently write to solicit a reunion and receive a polite rejection penned by Baker’s husband, Pepito Abatino (Birksted 2009, p. 345, note 143). Both episodes can be qualified as unsolicited attentions, paralleled by the black and white stripes, a coincidence in which some read Le Corbusier’s derisive parody to Loos’ project.

Loos' design for Baker house, however, has been the subject of extensive critical exercises across time, as telling of the relationship between Modernism and one of its most conspicuous muses.

Farès el-Dahdah's classical reading of Loos' project appeared on Assemblage 26, briefly retraces "a narrative history of an architecture complicit with a subject's affects" (El-Dahdah 1995, p. 73). El-Dahdah evokes a number of buildings structured as 'metaphors of the protagonist's love' starting with the oneiric ruins of the *Hypnerotomachia Poliphili* to end up focusing on *Le petit maison*, the architectural fantasy described by Jean-François de Bastide in his amorous novel of 1763.

The parallel between the unimplemented architectural compositions of *petit maison* and Baker House elaborates on several levels, but mainly on the correspondence of the building's distribution with the topography of the occupants' sentiments. However, while in the case of Bastide's novel the unfolding distribution of Marquis Trémicour's *petit maison* corresponds to the fluctuating topography of his guest Mélite's affects, in the case of the Baker House the architecture's composition ends up transubstantiating the desire of its own creator. Loos is, thus, trying to compensate for the absence of, or refusal from, the object of his colonial desire, Baker, by fantasizing her architectural transcription.

5.2.10 Baker Is the Landscape

El-Dahdah describes Loos' design as 'an *admirateur's* letter' (El-Dahdah 1995, p. 75), remarking the parallel with Le Corbusier's written *avances*. As an 'instrument of tactile extension', laden with no specific meaning but that of yearning, the amorous gift of the project is "a metonymic object to occupy the space where the other is not" (El-Dahdah 1995, p. 76). More than a machinery of seduction, it is, thus, a fetish prosthesis accompanied by Roland Barthes' formulation: "with this object, I touch you with my phallus". In fact, we could say that Baker is the landscape and Loos is trying to colonize her by architecture.

The metonymic character of the design is remarked also in the "flagrant script of Josephine's body" 'daubed' on the outside (El-Dahdah 1995, p. 77) in horizontal black stripes over the blank façades of Loos' architecture. The black stripes echo the livery of exotic zebras, macabrely flayed to line the walls with an *animalier* living tapestry. The black stripes also recall the horizontal line of the feminine body waiting to be penetrated by virile verticality. In *Ornament and Crime* Loos narrates that a primordial penetration informs the cross. In that, he retraces the primitive origin of decoration. The 'Papuan' tattoo, the primeval decoration of the aboriginal body, Loos' constant reference along his essays, daubs the striped exterior of the house with its mark of primitivism as it were transferred from the very skin of savage Baker.

Loos' construction is, however, an onanistic fantasy, an impotent's autoeroticism of which the colonial prey seems even unaware. He wants her to "wear the stairs" (El-Dahdah 1995, p. 79), he designs for her a grotesquely windowed pool to exhibit her nudity, as an exotic animal in a zoological cage or a nineteenth-century brothel's "*tableau vivant*" (El-Dahdah 1995, p. 79). But in private she probably aspires to sobriety as much as her public persona is exuberantly breaking the rules of civilization, since Loos' gift does not receive any documented reply, not even at the hands of Abatino.

5.2.11 Uneven Parallels

El-Dahdah's *petit maison*/Baker House parallel is, thus, quite uneven, for assimilating a bachelor's machinery of seduction with an onanist's apparatus of self-deception. However, the nucleus of the suggested correspondence between the XVIII century's *topos* of the *boudoir* within a *hotel particulier* and the pool construct in Baker House is equally relevant.

Both are specially intimate rooms, hidden in the interior of the architectural body, accessed from the sleeping quarters, and firmly "codified as a woman's territory" (El-Dahdah 1995, p. 80), where in some cases a lover can be allowed. Illumination is indirect and includes special effects of reflection, which in one case are produced by the mirror cladding of the walls, in the other by the protean multiple reflexes of water, whose suspended volume is pierced through by porthole windows. In both cases the real space is transfigured. The walls of Trémicour's *boudoir* are "covered with mirrors whose joinery was concealed by carefully sculpted, leafy tree trunks […], heavy with flowers and laden with chandeliers" (Bastide 1994, p. 75), so that it "could have been mistaken for natural woods" (Bastide 1994, p. 76). Rodolphe El-Khoury and Anthony Vidler retrace the architectural model of Bastide's literary fantasy in Jaques-François Blondel distributive and decorative lessons published in Paris in 1738 (El-Khoury 1994, p. 9; Vidler 1994, p. 19), starting a longer correspondence between the two.[15] The real manmade space of the *boudoir* is thus artfully converted into a fictitious natural environment. Similar is the case of Baker House's pool. In both cases a refined play of light dissolves the box of the room.

This expansion of a physical interior into a mental exterior, such as the endless space of Baker House's pool and *petit maison*'s *boudoir*, is a composition technique that would be largely adopted under different circumstances by Italian radicals starting with Archizoom's No-Stop City project of 1969. A miniature-like

[15]Along with El-Khoury, Claudia Conforti also notes that Bastide's account, initially published on a newspaper in 1760, might have a physical model in the Pavillion de Laboissière, or Hôtel de La Bouëxière, (Conforti 2017, p. 19) built for the tax collector Charles-François Gaillard de La Bouëxière, son of Jean, by Antoine Matthieu Le Carpentier in the mid-eighteenth century as a *folie* in the gardens designed by Jean-Michel Chevotet (Wikipédia 2018). Freely inspired by Bastide's original, Vittorio De Feo, the subject of Conforti's essay, had written a literary ecphrasis elaborated upon his own project for a little house, incarnating his lofty interpretation of architecture as a playful pleasure. See *Tre racconti di architettura* (De Feo 2010).

territorial module is endlessly multiplied by the reflections of the four mirroring sides of the containing box.

Commenting on the *petit maison* and its machinic furniture, Vidler attributes the dropping of the traditionally intended 'fiction of architecture' "in favor of an art of endless mechanical manipulation of space", among others, to the Marquise de Sade's *mise-en-scène* for the rituals of the *Cent-vigint jours de Sodome* (1785) (Vidler 1994, p. 15). The same erotic manipulation of real space is made more sophisticated and aware in *Superstudio*'s mirror chamber of the *Supersurface* installation at MOMA in 1972, in which the visitor is immersed. Manfredo Tafuri's accusations of senseless auto-eroticism and psychedelic delirium targeting the *neo-avant-garde*, are referred to the production of such fictitious spaces by means of a depraved manipulation of architecture, tending to replace the entire reality with 'communication' (Tafuri 1972, p. 394).

5.2.12 Re-Paradigmed Worlds

One Saturday afternoon, Danielle De Niese had just finished her performance at the New York Metropolitan Opera playing Despina in *Così Fan Tutte* to shed some light on Baker's seductions over the world of architecture.

My friend and music connoisseuse Miss O., insists that Susanna Phillips, the other soprano and actual *primadonna*, has a more powerful voice. But De Niese sings the Italian lyrics with a spell—I hardly catch a few words—that transmutes the language into an unknown birdsong flashing plumage all around.

Mozart and Da Ponte's opera buffa of 1790 is set in Naples and tells the story of two ladies from Ferrara engaged with young army officers. The daring young men improvidently accept a wager from Alfonso, an old cynical philosopher, who challenges the faithfulness of their dames. At the end of a series of amusing turnarounds, deftly managed by the young housemaid Despina, both ladies end up falling to the court of two mysterious suitors, whose noble Albanian disguise conceals no one else but their fiancées.

During the Glyndebourne Opera House Festival, De Niese accepts to shoot an interview on her unique voice warming up exercises. "Airflow—Trilling—Nasal Resonance… open up cheek muscles, loosen up tongue…", but with De Niese 'trilling' becomes a 'tongue-to-lip trilling', among a bunch of other variations. A Sri-Lanka Burgher of Dutch descent, when De Niese 'machines' her voice, she embodies 'the creole'. With reference to linguistic paradigms, the creole is the overcoming of the pidgin. Consequent to the relocation of European groups along colonial patterns in the 'New Worlds', the pidgin is a plane displaced fusion of two or more languages. The creole, instead, is the language of new natives, generations after that fusion. Not a language one can adopt by will, it is a cultural confluence in which one is born, that allows the subject to reformulate their horizon by re-paradigming new worlds, where the 'creole' is a more simply spelled substitute for 'miscegenation'.

5.2.13 *The Reign of Birds*

Deleuze Guattari discuss the "problem of the machining of the voice" as related to the "becoming-woman" and "becoming-child" of music, with reference to Dominique Fernandez' *Porporino ou les mystères de Naples*[16] (Grasset 1974), a novel on the life of a *castrato* opera singer. The concept of the 'machining of the voice' is specifically related to reformulations of opera singing modalities: with the 'head voice' the countertenor sings above his voice and his voice "operates inside the sinuses and at the back of the throat and the palate without relying on the diaphragm or passing through the bronchial tubes", while with the 'stomach voice' the *castrato* "gives carnal matter to the imperceptible, impalpable, and aereal" (Deleuze and Guattari 1987, p. 303).

Marcel Moré's *Le Dieu Mozart et le monde des oiseaux*[17] (Gallimard 1971) is a second reference to draw a parallel between Mozart's musical and vocal innovations and a process of "becoming-horse" and "becoming-bird" that characterizes the opera production in the second half of the XVIII century and the first three decades of the XIX century. Mozart's *accenti*[18] draw large diagonals transfiguring by analogy the horse's movement over the scores, while through 'embellishments' such as *gruppetti*, *appoggiature*, *staccati*, o *spiccati*[19] "the human musician is deterritorialized [that is transfigured] in the bird" (Deleuze and Guattari 1987, p. 304).

Mozart's *Così fan tutte*, composed little before his death, can probably be regarded as the epitome of the "becoming-bird" of baroque music. The deflagration of the 'embellishments' appears to be interpreted as potently by De Niese's 'tongue-to-lip trilling' as it had been by Porporino's androgynous voice in the 1770s. Porporino, Mozart, Baker, and De Niese are gods and demons of the 'reign of birds', who re-paradigm new worlds.

5.2.14 *The Reign of Insects*

In the '30s of the nineteenth century, it will be the rise of Verdi's and Wagner's voices, reterritorialized in the dual-gender system, to sanction the demise of the

[16]In English: Porporino or the Mysteries of Naples.

[17]In English: God Mozart and the World of Birds.

[18]An *accento* is an emphasis placed on individual notes.

[19]A *gruppetti*, *appoggiature*, *staccati*, *spiccati* are musical 'ornaments' not strictly necessary to carry the line of a melody. Subtracting from a principal note's time-value, an *appoggiatura* is an added higher or lower note. A *gruppetto*, or 'turn', is a figure composed about a principal note, assembling a higher note, the note itself, a lower note, and the note itself again. The *staccato*, is a sequence of notes of shortened time-value for each to be detached from the next, applying chiefly to wind instruments, while the *spiccato* is the technique of bouncing the bow on the strings to detach one note from the other.

Baroque musical wave, the oblivion of the androgynous option interpreted by *castrati* such as Porporino, and the early overshadowing of Rossini's and Bellini's modes (Deleuze and Guattari 1987, pp. 307–308).[20] According to Deleuze Guattari, despite their capitalistic and bourgeois drive, Verdi's and the Wagner's reformulations will eventually result in the diffusion of 'new molecular flows' leading to Debussy's 'molecularization of the motif', by virtue of their chorality and lyricism, and the consequent 'dissolution of form'. The reign of the voice is thus succeeded by the reign of the orchestra, where the voice is integrated.

The "reign of birds", vocal beings, is succeeded by the "reign of insects", instrumental beings, with "their vibrations, chirring, rustling, buzzing, clicking, scratching, and scraping", which seems to acoustically allude to the expansion of contemporary 'emergent' systems and diffusive occupation of the territory in the XX century.

5.2.15 Légion d'Honneur

The aura of Baker's persona would be publicly recognized after the end of the Second World War, when she is awarded the 'Croix de guerre avec palmes', the 'Rosette de la Resistance', and the 'Légion d'Honneur', after repeated obstructions from the military authorities under the mandate of French President Charles De Gaulle. Starting at the outbreak of the Spanish civil war, she had worked as an undercover agent for the French intelligence, digging secret information from Italian, Japanese, and German diplomatic sources. Braving out immediate execution, she had been transferring the documents hidden between her music sheets or in her panties, crossing the boundaries of Europe, South America, and Africa. Baker's civic figure would surprisingly consolidate, then, with her speech/performance delivered in her Free French Army uniform, in front of the Lincoln Memorial at the March on Washington for Jobs and Freedom of 1963, upon invitation from Martin Luther King.[21] Baker had self-reformulated herself twice, first from exploited juvenile to creole goddess, then from *grand diva magnifique* to icon of feminist, social, political, transnational, and racial emancipation, and twice her body had re-paradigmed new worlds.

---

[20]In fact, Rossini (1793–1868; *Il barbiere di Siviglia*, 1816, prequel of Mozart's *Le nozze di Figaro*, from the homonymous dramas of 1775 and 1784 by Pierre-Augustin Caron de Beaumarchais) was active until the fourth decade of the 1800s, while Bellini (1801–1835; *Norma*, 1831) did not survive the early triumphs of Verdi (1813–1901; active 1839–1893; *Nabucco*, 1842) and Wagner (1813–1883; active 1832–1882; *Der Ring des Nibelungen* tetralogy, 1869–83).

[21]Still in 1936, Baker had been called 'negro wench' by the New York Times upon her unacclaimed return to America, with the intention of relegating her to the censured space of segregation. Several black activists would in turn regard her artistic roles as degrading for the Afro-American community.

5.3 'Miscegenated' Spaces of Nueva España

A review of the typical arrangement of Nuevo-Hispanic public space grounds abstract notions of 'miscegenation' into practical terms of our discipline. The process of hybridization between the Mesoamerican open space model and the type of the Euro-Mediterranean plaza, overlaid throughout the Spanish conquest, (Wagner et al. 2013) is the spatial correspondent of the equivalent process of cultural 'miscegenation' described by Bernal and, graciously, precipitating in Baker's glittering string of bananas.

The Mesoamerican open space is an uncontained liquid expanse, a material referent of the 'primordial sea' of genealogic myths, identified by a distribution of masses, sculptural pieces or buildings that transfigure characteristic arrangements of rocks, mountains, or other natural features. It is a manmade transcription of surrounding landscape assemblages. The type of the Euro-Mediterranean plaza is a contained zone, in which the traditions of the Greco-Roman agora-forum, the medieval market place, the Vitruvian ideals reinvented by the Renaissance, and the Islamic *maidan* have confused.

The spatial organization at the core of the colonial cities founded by the Spaniards in *Neuva España* during the sixteenth and seventeenth centuries, always imply a composite assemblage that, intentionally or not, hybridizes the type of the Euro-Mediterranean plaza with an openness characteristic of Mesoamerican arrangements. An exemplary case is that of the *zócalo* of Oaxaca, where the central public square is flanked north by the side of the cathedral occupying large part of an

Fig. 5.4 Public space system in the historical center of Oaxaca, Mexico, with zocalo square, cathedral, atrium, and alameda within Hippodamian grid (*photo* images ©2018 Digital Globe, Map Data ©2018 Google INEGI Mexico)

urban block. A residual C-shaped area left open around the *pronaos* of the cathedral functions as the sacred *atrium* while retaining a civic use. The *alameda*, a rectangular plaza densely planted with trees attached to the west side of the *atrium*, reinforces an east–west axis perpendicular to the north–south axis of the main square. The result is an L-shaped urban ensemble formed by the open spaces of the *zócalo* and *atrium-alameda* jointed by the cathedral. The *zócalo* and the *atrium-alameda* are fluidly connected on a southeast to northwest diagonal (Fig. 5.4). The Oaxacan *zócalo* area is a 'miscegnated' urban assemblage resulting from the hybridization of spatial types of diverse origins. The same spatial organization, due to intentional and unconscious instances of the collective memory of miscegenation, governs the urban ensembles of the *zócalos* of the *centro histórico* and Coyoacán in Mexico City. The former case shows a subtler stratification of the open space system, articulated in plaza, cathedral, and Templo Mayor—the last term being a latent pre-Hispanic presence, slightly recessed from the manifest composition of the Hispanic complex.

A similar process of miscegenation is represented by Nuevo-Hispanic types such as the enclosed courtyard, combining Mesoamerican triad-centerings, quincunxes, and, quadrangles with Mediterranean cloisters, *horti conclusi*, and madrasas.

5.3.1 Expropriations in the New World

We have discussed in Chap. 2 of how the *locus* can be interpreted as the confluence of history and geography, where trans-historical aspirations are developed by an identitarian group dwelling across time on a region. The territorial dynamics carried forth in France between 1789 and 1871, referred to as 'land expropriations', show surprising analogies with the expropriation policies enacted and fragmentarily implemented in Mexico between 1786 and 1857. In the former case, we can consider the outbreak of the French Revolution and the carnage of the Paris Commune as the precipitating points of a broader historical wave. In the latter case, the Bourbons' Real Ordenanza de Intendentes for colonial Nueva España and the Leyes de Reforma starting with the liberal constitution for independent Mexico provide the reference.

As Pöete's and Halbwachs' analyses describe how the historical sequence *ancien régime*-French Revolution-Empires[22] grounds itself into the urban area of Paris, in his substantial monograph of 1997 about collective identities in Mexico, Enrique Florescano brings together a broad historical fresco of the transient regimes unfolding from the ancient Mesoamerican civilizations to the Porfiriato via

[22]We can identify a complete historical sequence by lining up: *ancien régime*; 1789 revolution; 1792 terror; 1794 white terror; 1801 First Empire; 1815 Restauration; 1830 July Monarchy; 1848s Republic; 1851s Empire; 1870 Third Republic with the 1871 parenthesis of the Communes.

Virreinato, Independencia, and Reforma.[23] By analyzing the implemented and aborted reforms on the structure of the immense Mexican nation, with its endless geographic and ethnic diversity, Florescano sheds light over a miscegenated accumulation of layered fragments and traces. Alongside the expanded sociological narrative, collateral frames of the century-long region-forming can be retraced.

5.3.2 *Merciless Dissection of a Country's Destiny*

Florescano's comprehensive historical reconstruction in *Etnía, Estado y Nación*[24] proposes to contribute to an unbiased understanding of the *campesino* rebellion breaking out across Maya lands under the flags of the Zapatista movement in 1994. The focus of the analysis is on the dynamics involving the collective identity of the dwelling groups, the lands of the fathers, and the structures for their government. Florescano is often assimilated to 'subaltern historiography' for his interest in the perspective of native communities and refusal of both post-colonial nationalism and neo-Marxist social readings.[25] However, compared to other subaltern histories, his work exhibits a more detached gaze over the human events, by neither idealizing the pre-Colombian nor the colonized native. He seems to observe, rather than condemn, violence and conflicts as agents of the historical unfolding. For this very character, his anti-third-worldist vision can be considered compatible with both Sergio Romano's nostalgic reconstruction of the colonial process of territorialization in *La città europea fuori d'Europa* that we will touch upon further on in this chapter, as well as Pöete-Halbwach-Rossi's model of interpretation of spatial transformations based on Engelsian dialectic materialism described in Chap. 1.

Other authors before Florescano have reserved darker tones to the *estado nacional* of the Reforma compared to the colonial regime of Nueva España,

---

[23]A complete historical sequence, in this case, lines up: Mesoamerican civilizations; 1325 foundation Tenochtitlán future capital of the Aztec Empire; 1519 Conquista; 1521 colonial Virreinato de Nueva España; 1810 war of Independence; 1821 Independencia; 1822 Imperio de Iturbide; 1823 United Mexican States; 1824 Republican Constitution and Primera República Federal Mexicana, 1836 centralized reform of Santa Anna and dictatorship; 1846 US Mexico war; 1846 Segunda República Federal Mexicana; 1847 Yucatan's *guerra de castas* start; 1848 loss of a half of the territory to the USA; 1854 Plan of Ayutla; 1855 liberal government; 1857 liberal constitution; 1858s dictatorship of Santa Anna; 1859 Leyes de Reforma; 1861 liberal government of Juarez; 1864 French intervention and Second Empire; 1867 Liberal Government of Juárez; 1872 Liberal Governemt; 1876 Porfiriato; 1810 Mexican Revolution; 1917 Constitution and civil war; 1929 Maximato; 1934 PRI one-party rule.

[24]In English: Ethnicity, State, and Nation.

[25]Florescano intends to override the politically interested reconstructions of the 1994 events and their motives: on one hand, the perpetuations of the positivist nationalism that has kept the country under authoritarian rule, excluding the authentic participation of native groups; on the other hand, the celebrative salutations from the neo-Marxist *intellighencija*, which subject the interpretation of the historical dynamics to alien ideological superstructures.

dismantling the edifice of a conventional narrative (Hale 1968, 1989). Others, such as Jacques Lafaye in his classical book *Quetzacoatl and Guadalupe*, have previously described the ethnic miscegenation of Pre-Columbian and Marian components, which would deliver the Mexican nation through the transition from *colonia* to *estado nacional* (Lafaye 1976). Florescano's offhand synthesis covering the span of two millennia has gained him conflicting critiques about the bending of historical factuality in favor of a vision. What makes Florescano's work particularly relevant to this research, however, is the fact that it can be used to relate the trajectories of the ethnic identities in Mexico to the fluctuations of the territorial structures of the country that is its spatial transformations.[26]

Florescano's analysis is organized in the three historical phases of the pre-Columbian era, the colonial society of the sixteenth–eighteenth centuries, and the nineteenth-century national state, to conclude that the discontinuities in the Mexican sociological regime—that we hereby try to interpret for the consequent territorial shifts—occurring with the advent of the national state are, for many, even more dramatic than those descending from the advent of the colony. Despite exerted on the past, the work is a merciless dissection of Mexico's destiny projected into the country's future.

5.3.3 Altépetl: Sowing, Monuments, Land Rights

When the bands of hunters-gatherers start their process of independization from nature by settling to cultivate and cattle in the Mesoamerican cone, the first permanent centers are formed. By 1200 AC, the Olmec culture has edified the 1-km-long monumental terrace of San Lorenzo (in today's state of Veracruz) and, from there, irradiates towards the territories of the present-day Mexico valley, Morelos, Guerrero, Oaxaca. In a later phase, the Olmecs implement a series of pyramidal constructions of rammed earth flanked by flooded patios on islets in the Río Tonalá. While the ponds correspond to the primordial ocean, the tri-stepped pyramids are the ceremonial correspondent of the primigenial mountain, stacking

[26]Ethnicity is defined as a group of individuals historically settled in a territory, with a common language and culture, who recognize their own peculiarities before other groups and identify themselves by an own name. The state appears with the legal system intended to exert sovereign power over a territory and the individuals in its jurisdiction. The nation, which anciently coincided with the ethnic group, is reformulated into the new idea of citizenship by the overturns of the French Revolution. Since then, the members of a nation reciprocally recognize themselves mutual obligations and rights by virtue of their common belonging to a community of peers, that is not by virtue of birth, but social edification of the citizen. To formulate the notions of ethnicity, state, and nation, Florescano relies respectively on Tamara Dragadze, Colin Renfrew, Anthony Smith; Norberto Bobbio; Eric Hobsbawm. This ideal of a modern nation of citizens, governed by equal laws, corresponds to an "imagined community" (after Benedict Anderson), in which the inhabitants of the country, with all their disparities, are unified by similar ideals, share a territory, have a common past, and venerate symbols of identity (Florescano 2001, pp. 14–16).

up underworld, earth, and sky along the cosmic axis. According to Florescano, "al sembrar la tierra de cultivos y colmarla de monumentos, los pobladores adquirieron un 'derecho' de propiedad sobre ella; la tierra se convirtió en territorio de la comunidad y se vinculó a los antepasados y los dioses protectores" [27] (Florescano 2001, p. 31).

These basic elements will be recurrent in all the Mesoamerican cultures. The pyramid is the artificial equivalent for the surrounding mountains, set onto an expanse of flat land representing the originary waters. The monumentality of the complex is sanctioned by the ancestral rites perpetuating the memory of the group's origins. The primordial myth is thus implanted into the territory along the vertical axis underworld-earth-sky. This structure is immortalized in numerous Mesoamerican bas-reliefs, representing the divine ancestors in the act of surfacing from the underworld through the sacred cave at the foot of the water-filled mountain with the seeds of maize in hand.[28] The seeds will be scattered over the territory to generate common people as well as crops, while the nobles are direct offspring of the ancestors or variously born from tree trunks or aquatic turtle carapaces. The town-territory identity in Mesoamerican cultures is a strong reminder of the agricultural origin of the city and its bounds to the countryside. Water-filled mountain equals chief town equals territory equals people equals crops, by virtue of the ethnic memory deposited in mythology and vertically organized along the cosmic axis.

In Nahuatl, the word '*altépetl*' summarizes this composite notion. Made up of '*atl*' meaning water and '*tepétl*' meaning mountain, the 'mountain filled with water' is the term that refers to a state as a whole territory subjected to the rule of an ethnically determined group settled in the chief town, as we will discuss more in detail further on.

5.3.4 Vertical Axis: Ethnicity, Territory, Cosmos

The Zapotecs start consolidating their villages in the Oaxaca valley since the seventeenth-century BC. By the twelfth-century BC, the chief centers gathering the resources of the subjected territory are ruled by *caciques* of noble lineage. In the structure of the *cacicazgos*, the condition of equality among the members of the nomadic band is broken and the hereditary character of power is sanctioned.[29] The Zapotec culture exhibits its climax between the first-century BC and the

[27]In English: "by sowing crops in the land and filling it up with monuments, the dwellers acquired a property right on it; the land became territory of the community and was bound to the ancestors and patron gods" [translated by the author].

[28]See the San Lorenzo basalt throne, the toponymical glyph of Mount Tláloc in the Códice Borbónico, and Olmec steles, murals of Teotihuacán, Oaxacan bas-reliefs, Cholulan drawings (Florescano 2001, pp. 36, 268, 270).

[29]Cultures based on the *cacicazgo* regime settles between the sixteenth- and the thirteenth-century BC in the area of Soconusco, in the Isthmus of Tehuantepec.

Fig. 5.5 Acropolis of Danibaan (today, Monte Albán), Oaxaca. The vertical apparatus of the cosmic axis in its conspicuous manifestation. On the background, the built masses/orography correspondence (*photo* rp)

second-century AC, immortalizing the structure of its state in the monumental acropolis of its capital Danibaan, today Monte Albán (Florescano 2001, p. 29) (Fig. 5.5).[30]

The Maya cultures flourish between the fourth and the tenth century in the areas of Chiapas, Campeche, Yucatán, Belize, and Guatemala, imposing an urban layout to their confederated centers, such as the prominent states of Tikal and Kalak'mul, with a formalized plaza surrounded by pyramidal monuments. In plazas such as the one of B'aakal (today, Palenque), the natural and the architectural components are combined in a grand scenery for the ritual representation of power and the identification of the governors with the divine ancestors,[31] the state's territory with the Earth, and the city with the cosmos (Florescano 2001, pp. 49–50, 53, 60, 63). The

[30]Danibaam will decay by the eighth century.

[31]See the famous bas-reliefs from Palenque representing Kan Balam's ascension and the transfer of power and ritual tools from his father Pakal in the seventh century.

ethnic principle is therefore anchored onto the territory and aligned with the cosmos by the vertical axis, the spatial manifestation of the myth of the origin.[32]

5.3.5 *Permanence of the Collective Character*

Teotihuacán in the Mexico valley is the first settlement to pursue and achieve the concentration of a substantial part of its people within the urban precincts. The construction of the monumental axis of the 'Calzada de los Muertos' in the first century will remain the major feature of the settlement throughout the evolution of the city state until its destruction in 750 AC. The second phase of the evolution of the city starts around 300 AC with the transition from the traditional monarchic rule to a regime centered on a set of religious collective values. The further development of the city is implemented on an ambitious plan, for the first time in Mesoamerican history extended to the residential areas. Large urban streets and blocks are meant to receive the entire population orbiting around the city, including the farmers (Florescano 2001, pp. 66–72).[33] However, the development is anchored to the ancient monumental complex of the political-religious Citadel and to its myths. The Pyramid of the Sun covers the cave of the ancestral myth of creation, seat of the Goddess of the Waters, while the Pyramid of the Moon carries the temple of Tlalóc, the God of the Storms, and at the north end of the Calzada is set the temple of Quetzacoátl.

Teotihuacán and the emblem of the Plumed Serpent become the image of power in the Post-classical period expanding its influence as far as the Maya regions in the south and the lands of the nomadic hunters in the north. The post-classical Maya city of Chichén Itzá erects its grandiose monumental complexes in the ninth and tenth centuries exhibiting evident Toltec influences. The plumed serpent, in fact, descends the steps of the Kukulkán pyramid as it decorates the government palace of Tula. The collective character of the regime, like in Teotihuacán, results conspicuous from the bas-reliefs apparatus (Florescano 2001, pp. 87–9).

When the frontier between the bands of nomadic hunters of the north and the sedentary farmers of the plateau disappears in the seventh century, groups of Chichemecs descend into the Mexico valley, merge with groups from the Teotihuacán culture and settle in Tula,[34] or Tollán in Nauhatl, with the name of Toltecs. Tula, the legendary place of the origin in the Aztec mythology, is described as a grandiose center irradiating art and culture, quite in contrast with the modest archeological evidence. Tula's assimilation with the greatness of the ancient Teotihuacán could be explained by the presence of prevailing cultural components

[32]The state of B'aakam, along with the Maya cultures, decays in the 9th century.

[33]The city trace included even a district exclusively inhabited by Maya communities, which has led to consider Teotihuacán the first multiethnic state of Mesoamerica.

[34]Ah Puh in Maya.

from the old capital (Florescano 2001, p. 80). By that, the Aztecs are also acknowledging the collective regime of the ancestral state.

5.3.6 *Mythic Peregrinations from Nomadic to Sedentary*

A characteristic trait of the myth of the origin in Mesoamerican cultures is that of the peregrination, which the Aztecs formalize in their ancestral codices. In the Códice Boturini,[35] the fatherland on the island of Aztlán becomes a golden age myth, forged as a specular construct on the image of their homeland of Mexico-Tenochtitlán, built on the islet amid the Texcoco lake after their century-long and perilous migration. The Codex Xólotl, a 16th century copy of an older pictographic manuscript,[36] shows the migrations of Chichimec tribes led by warrior-leaders from the north into the region east of the Texcoco lake.[37] The miscegenation between Chichimec groups and sedentary tribes is vividly illustrated by the Mapa Quinatzin,[38] another sixteenth-century pictorial document in three sheets recording the history of the region. The first sheet shows barbaric Chichimecs, covered with furs and carrying bow and arrows, meet civilized Acolhuas, dressed in fine textiles and devoted to the care of their intensively cultivated territory. In the merging, the barbarians apprehend the engineering of irrigation and the Nahua language while granting fierce courage and military prowess.

While the island of Aztlán is clouded in the fogs of the legend, the March of the Aztecs from Tula to the Texcoco basin can be historically reconstructed (Florescano 2001, pp. 98–102). Their migration can, in fact, be considered a peregrination. The Aztecs are purportedly following an idol carried at the head of the procession, representing Huitzilopochtli, their god associated with the sun, and his prophetic

---

[35]See the Códice Boturini or Tira de la Peregrinación, Biblioteca Nacional de Antropología e Historia de México, amatl paper (*ficus* bark) with Mexica writing, 21-folds 20 × 542 cm, conventionally dated 1530/41.

[36]Codex Xólotl, Bibliothèque nationale de France, Département des Manuscrits, Mexicain, amatl paper (*ficus* bark), 10 folios 40 × 50 cm, conventionally dated 16th century. The document accounts the history of the Acolhuas, from the arrival of the Chichimec tribes into the area of the Texcoco lake (first folio) until the conquest of Azcapotzalco, on the west shore of the lake, at the hand of Netzahualcóyotl of 1428 (tenth folio).

[37]The lake is named by the capital of the Acolhuacán state, Texcoco.

[38]Mapa Quinatzin, Bibliothèque nationale de France, Département des Manuscrits, Mexicain, amatl paper (*ficus* bark), 3 folios 34.5 × 43.5 cm, conventionally dated 1543/48. About this document see the detailed and valuable analysis by María Mohar (Mohar 1999). We refer in particular to the first folio illustrating the encounter between the new Chichimec peoples and the native dwellers of Acolhuacán and aspects of their life. The second folio describes the palace of Netzahualcoyótl, emperor of Texcoco, and the third crimes and relative punishments of Acolhua legal system. Both the Codex Xólotl and the Mapa Quinatzin were part of the collection of the 'erudite chronicler of the Acolhuas', Fernando de Alva Ixtlilxochitl (1568/80–1648).

prediction of a new homeland. While marching, they change their name to Mexicas to acknowledge the profound transformation of state their people is undergoing.

Upon reaching the western shores of the Texcoco lake, the Mexicas try to settle in various locations, including the cliff of Chapultepec, being expelled by the groups settled in the area. They are pushed to take refuge onto an islet (*tetl*) hidden by bulrushes at the center of the lake, where they see an eagle poised upon a nopal intent on eating its prickly pears (*nochtli*), as it had been predicted to them. There, they found their capital of the Mexicas, México-Tenochtitlán (*tetl-nochtli-tlan*, or place where prickly pears are found on an islet) in 1325 or 1345 (Florescano 2001, pp. 101, 119).[39] The legendary narration of the peregrination from Aztlán, can be read as the way in which the Aztecs envision their own process of becoming sedentary.

5.3.7 A Water-Filled Mountain and an Empire

By lending their military services to the Tepanecs of Atzcapotzalco, settled on the western region of the lake, the Mexicas can start the consolidation of their state in the shadow of a dominating nation. Through a policy of strategic marriages, they also interlace alliance with the opposed front of the state of Texcoco based on the eastern shore of the lake. Finally, profiting from an internal feud for the succession to the throne of Atzcapotzalco,[40] the Mexicas rout the Tepanecs in alliance with the Acolhuas in 1428. The imperial expansion of the Mexicas can start under the leadership of their fourth *tlatoani*, Itzacóatl. It will lead them to put together a vast territorial entity expanding over the Mexican plateau from the Atlantic to the Pacific bordering with the lands of the Chichimecs north and the lands of the Zapothecs south.

The Aztec empire is organized as a system of tributary territories formally ruled by an alliance of three cities, the so-called Triple Alliance, but in practice hegemonized by Tenochtitlán. The empire even includes areas detached from the main entity, such as Soconusco in the Isthmus of Tehuanepéc, which in fashion of colonies supplies the capital with exotic cargos of the most exquisite cacao. The structure of the Aztec empire, however, is quite different from that of European models. It is organized on the base unit of the *altépetl*, the water-filled mountain, that is the city-territory that the Mexicas had adopted from the sedentary groups of

[39]See also Códice Mendoza, Bodleian Library, University of Oxford, 71 folios with later explication notes in Spanish, dating back to the 1540s, three sections: listing the Mexica governors and their conquests from 1325 to 1521; showing the *altepemeh* under Mexica rule and their tributes; daily aspects of the life of the Mexicas. See folio 2 illustrating the foundation of Tenochtitlán.

[40]First, they manage to re-establish their ally Nezahualcóyotl on the throne of Texcoco and then attack Azcapotzalco with the alliance of the *atépemeh* of Texcoco and Tlacopam, then known as Triple Alliance, base of their future empire.

the plateau, also common, as mentioned before, among the Maya peoples of the south.[41]

The *altépetl* is a 'territorial unit' on which 'one or more ethnic groups with a common past and traditions' are dwelling, governed by a 'dynastic lord' called *tlatoani*. At the center of the *altépetl* rises a temple, both 'house to the tutelary god' and 'symbol of the territorial sovereignty' concentrated in the primordial mountain, and a monumental plaza is the 'ceremonial and market center'. From the plaza outwards, the districts are laid-out at the four cardinal points (Florescano 2001, pp. 267–8).[42] Urban and rural lands irradiate from the chief town of the *altépetl* reaching the border where other *altepemeh*[43] come together to cover the territory in a net of interlaced patches. The interlacing of the patches is nested in descending levels of subordination from the capital to minor commons.

5.3.8 *Of Massacres and Foundations of Cities*

We have already mentioned of the landing of Cortés on Maya shores in 1519. In *La cittá europea fuori d'Europa* (Benevolo and Romano 1998), the magnificent anthology that interprets the European colonization of the New World through the urban foundations imposed onto it, Romano epitomizes the colonial history by aligning a series of figures that perpetuates it: Hernán Cortés, Bernal Díaz del Castillo, Francisco Pizarro, Diego de Almagro, Rodrigo de Bastidas, Vasco Nuñez de Balboa, Diego de Velásquez, Alvaro Nuñez Cabeza de Vaca, Hernando de Soto…

> "questi e altri sono gli uomini che conquistarono un continente. […] Ebbero molti vizi. Massacrarono gli indios, ne tradirono i capi, si arricchirono senza pudore […]. Ma ebbero anche straordinarie virtù. Furono soldati, esploratori, amministratori, fondatori di città e organizzatori di provincie. A dispetto di certe versioni bigotte e politicamente corrette della cultura terzomondista degli ultimi anni la conquista spagnola del Nuovo Mondo fu una delle più straordinarie pagine della storia d'Europa".[44] (Romano, Dopo il trattato di Tordesillas: l'emisfero spagnolo (introduction to Part III), 1998, pp. 61–2)

[41]While Zapotecs and Mayas prefer to represent their states by the toponymical glyph of the capital, the Toltecs and Mexicas represent their territory by identifying its boundaries. The emblems of the surrounding states are aligned all along the territorial limits in a diagrammatic mapping that also traces the center of the state and the principal paths, places, and rivers.

[42]Florescano bases his description of the *altépetl* on James Lockhart's sociopolitical studies on the Nauha communities in the sixteenth and seventeenth centuries.

[43]Plural of *altépetl*.

[44]In English: "these and others are the men who conquered a continent […]. They had many vices. They massacred the natives, betrayed their chieftains, enriched themselves shamelessly […]. But they also had extraordinary virtues. They were soldiers, explorers, administrators, founders of cities, and organizers of provinces. Despite certain bigot and politically correct versions of the third-worldist culture of recent years, the Spanish conquest of the New World was one of the most extraordinary pages of Europe's history." [translation by the author]

It is fair to mitigate Romano's sharp claims by also recalling the alternative 'vision of the vanquished', after the title of Miguel León-Portilla's celebrated collection of Nahua sources prior to or immediately following the Spanish conquest of Mexico (León-Portilla 2008 [1959]), that is the vision of the conquest from the point of view of the conquered. León-Portilla's anthology of native voices presents several excerpts from Fray Bernardino de Sahagún's *Historia general de las cosas de la Nueva España*. The *Historia* has been qualified as the result of a precursory anthropological research, conducted by the fray through direct inquiries among the elders of native communities and reports collected by his *informantes*, his former students at the Colegio de Santa Cruz Tlatelolco members of the Nahua elite, a class of trilingual intellectuals equally in control of Nahuatl, Spanish, and Latin. Sahagún's monumental work, relentlessly compiled and edited both in Nahuatl and Spanish across a period of about thirty years until 1577, is officially aimed at understanding the cultural context of Nueva España to facilitate its evangelization by the action of the Catholic missionaries. In the process, the author inevitably falls fascinated by a sophisticated culture at the brink of annihilation whose traits the work ends up perpetuating. The *Historia* presents two parallel texts, opposing a complete version in Nahuatl, transliterated in the Latin alphabet, to an abridged Spanish translation. Despite Sahagún's astutely orchestrated discrepancies, retaining the sharper comments from a native perspective within the Nahuatl text, the *Historia* would fall into the century-long oblivion of inaccessible archives decreed by Spanish censorship.

The narration of the carnage of Cholula in the translation from the Nahuatl text by León-Portilla's mentor, Ángel María Garibay, presents the voice of the natives with shaking intensity. In *'la matanza de Cholula'*, thousands of armless inhabitants are massacred at the order of Cortés, with the sole purpose, according to the Cholultec version, of intimidating the Aztec emperor in Tenochtitlan and harnessing the discords among native peoples[45]

> "Hubo reunión en el atrio del dios. Pues cuando todos se hubieron reunido, luego se cerraron las entradas: por todos los sitios donde había entrada. En el momento hay acuchillamiento, hay muertes, hay golpes. ¡Nada en su corazón temían los de Cholula! No con espadas, no con escudos hicieron frente a los españoles. [...] Y en tanto que todo esto se hacía, todo se le hacía llegar, se le decía, se le hacía oír a Motecuhzoma. [...] Es como si la tierra temblara, como si la tierra girara en torno de los ojos." [46]

---

[45]Following the Cholultec account presented here, León-Portilla also reports the excusatory Tlaxcaltec version accusing the Cholulas of being the cause of the massacre for refusing to submit to the Spaniards and killing a Tlaxcaltec emissary. According to the Spanish accounts of the events, instead, the Spaniards act to foil a lethal complot being prepared by the Cholulas.

[46]From Garibay's Spanish version presented in León-Portilla's *Visión de los vencidos: Relaciones indígenas de la conquista* of 1959. In English: "Assembly had been called in the temple of Quetzalcoatl. As soon as all had gathered, the entrances were taken and death was given by sudden stabbings and hits. Nothing the Cholultecs had feared in their hearts, no swords, no shields had they carried. [...] All the turmoil was reported to Motecuhzoma at once, by messengers running back and forth on the routes. [...] It was as though the earth trembled and swirled before one's eyes." [translation by the author]

Lysander Kemp's English translation is as intense as Garibay's version

> When the massacre at Cholula was complete, the strangers set out again towards the City of Mexico. They came in battle array, as conquerors, and the dust rose in whirlwinds on the roads. Their spears glinted in the sun, and their pennons fluttered like bats. [...] Some of them were dressed in glistering iron from head to foot; they terrified everyone who saw them. Their dogs came with them, running ahead of the column. They raised their muzzles high; they lifted their muzzles to the wind. They raced on before with saliva dripping from their jaws.[47]

The massacre is followed by the ruthless pillaging of one of the largest cities in Mesoamerica and the destruction of its pyramidal complex.

5.3.9 Territorial Continuity in the Age of the Vice-reign

Despite crossing the Atlantic in opposite directions, Romano's and Florescano's voices resonate. Florescano's distinctive idea, for the sake of our thesis, induced from Charles Gibson's and James Lockhart's sociopolitical analyses (Gibson 1995; Lockhart 1992), is that the territorial reorganization of Nueva España following the Spanish conquest pivots on the adoption of the *altépetl* as the base unit and its network as the bearing structure as "the cell upon which the institutions that organized the life of the native villages during the Vice-reign were built: first the *encomienda*, then the *distrito parroquial* and later the *cabildo español*"[48] (Florescano 2001, p. 271).

With the Vice-reign, the Spanish *encomenderos* collect a large part of the tributes leaving a minor portion to the native *cacique*, that is a *tlatoani* divested of the administration of government and justice, while the *altépetl* retains its territorial and social structure (Florescano 2001, p. 272). After all, the *encomienda*, introduced by Cortés following the take of Tenochtitlán, is probably the most effective policy to control and exploit a nested system of territorial levels with over 10 million subjects,[49] abruptly taken over by a coup led against its head by a contingent of a few thousand aliens. Cynical and pitiless, the *encomienda* relies on the conveniences of the privileged among localized ethnic groups, and consequently on the existing structure. Eventually, the *encomienda* installs parasite agents at the head of the existing territorial units.

The *repúblicas de indios*, sort of rural municipalities surrounding the newly founded colonial cities governed by Spanish, soon replace the *encomiendas*, considered an inhuman regime leading to individual accumulation of disproportioned

---

[47]From Lysander Kemp's English translation of 1962. See *The Broken Spears: The Aztec Account of the Conquest of Mexico*, Boston: Beacon Press.

[48]Translation by the author.

[49]The native population of Nueva España would suddenly drop down to 1 million, due to well-known epidemics by the end of the sixteenth century to then rise to 2.5 million at the end of the colonial era.

riches. The system of the *repúblicas de indios* allows for ambits of autonomous administration by part of the native groups, and is supported by a simultaneous action of cultural permeation operated by the church through the correspondent religious jurisdictions of the *parroquias indígenas*. *Repúblicas* and *parroquias* still coincide, by and large, with the old territorial structure of the *altépetl*.

The rediscovery, formalization, or fabrication of the *títulos primordiales* of the local communities is a distinctive phenomenon of the colonial regime of Nueva España in the seventeenth and eighteenth century until the end of the colonial era. The documents, written in Spanish or native languages, claim the rights of a community over a territory with particular emphasis on the communal lands. The 'titles' are customarily constituted by virtue of a historical reconstruction of the state of land jurisdiction and property in pre-Hispanic times, and subsequently sanctioned by the new Spanish colonial authorities. Often, the figures of pre-Hispanic ancestors of the community and first Spanish authorities overlap, (Florescano 2001, p. 277) so that the collective memory of before and after the conquest results relayered onto a territorial structure that remains largely permanent.

5.3.10 Territorial Discontinuities in the Age of Enlightenment

However, after the first violent wave of the colonization, the Spanish Crown introduces rebalancing measures that grant some sort of tutelage and some margins of autonomous administration in the rural areas. The grievances and denunciations of Fray Bartolomé de Las Casas, a former conqueror who had renounced to his *encomiendas*, result in the 1542 promulgation of the Leyes Nuevas prohibiting the enslavement of natives at hand of *encomenderos* under the tutelage of the crown and the surveillance of the friars. De Las Casas will then serve as bishop of Chiapas and Universal Protector of All the Indigenous. The natives are kept in a permanent state of tutored minority. To this respect, Florescano asserts that the idea of a colonial government as a monolithically repressive power onto the natives is an 'ideological creation' of the nineteenth-century creole historians who are edifying the narrative of the national state (Florescano 2001, p. 241).

A major disruption of the settled territorial regime occurs in coincidence with the wind of the Enlightenment blowing from Europe, which announces itself with the Real Ordenanza de Intendentes of 1786 emanated by the modernizing rule of the Bourbons. The Enlightenment, led by the faith in the rights of the individual, counters the corporative forms of medieval origin as well as the ancestral rights on communal properties. The *ordenanza* establishes functionaries to oversee the administration of the communal funds and lands of native communities.

Its consequences are the loss of the management of the collective patrimony of the native communities and the rights of access to woods, grazes, quarries, aquifers, hunting-gathering grounds. The implementation of such measures has the effect of coalescing the natives with the *criollos* and *mestizos*[50] in an opposing front at the turn of the eighteenth and nineteenth centuries (Florescano 2001, p. 224).

The ethnic diversity in the Spanish colonies had become very articulated during the previous two centuries. European colonizers, Mesoamerican natives, African slaves deported to work on the plantations, and later also the Filipinos and Chinese, immigrated as cheap workforce on the west coast, would interbreed generating a nested series of mixes. The colonial society would try to organize the results of the miscegenation of the three major groups, Spanish, blacks, and indigenous, in a system called *castas*, that is 'races'. It is a deterministic system of attribution of each American inhabitant to a specific ethnic group, on the basis of their genetic ascendency, and a correspondent social role, but does not prevent contiguity and intercourse. The *cuadros de castas* is a painting genre that develops during the eighteenth century to offer an articulated panorama of the racial diversity of the Spanish colony to the European elite on the other side of the Atlantic. The first systematic *castas* series is commissioned by the viceroy Fernando de Alencastre Noroña y Silva from Juan Rodríguez Juárez to show king Felipe V the varied racial miscegenation of Nueva España at the beginning of the eighteenth century.[51]

However, the contiguity among distinct ethnic groups within the social ensemble, where each part is acknowledged certain roles and rights, is a corporative construct on which medieval Europe and pre-Hispanic order overlap. The enlightened drop of social corporativism marks the discontinuity that will also affect the territorial regime.

5.3.11 Restructuring the Role of the Individual

The winds of the Enlightenment bring a profound restructuring of the role of the individual within the social body, ideally transitioning from a system of inequality

[50]By the name of *criollos* went the ones born in the colonies of pure European descent, while the *mestizos* are the ones born to mixed couples, generally European and Mesoamerican native.

[51]The Galería de Castas Mexicanas of the Lydia Sada de González collection at the Museo de Historia Mexicana of Monterrey, among other *pinturas de castas*, includes a typical 16-canvas series by the Pueblan painter José Joaquín Magón where the racial mixes of the three major groups, Spanish, black, and indigenous, are associated to standardized physical traits and psychological and personal characteristics: Español + India = Mestizo (humble, quiet, simple); Mestizo + Española = Castizo (attachment to horse); Español + Castiza = Fructo bello (alike his father); Blanco + Negra = Mulato (pride, disdain); Español + Mulata = Morisco (to be and doctrine); Español + Morisca = Albino (short-sighted, weak, soft, benign); Albino + Española = Torna atrás (figure, genius, customs); Mulato + India = Calpa Mulato (indocile, strong, stocky body); India + Calpamulato = Gibaro (restless, arrogant); Negro + India = Lobo, mala ralea (squanderer, hustler); Lobo + India = Cambujo (heavy, indolent, dull); Indio + Cambuja = Sambayga (the one for whom there is no guile he doesn't understand); Mestizo + Mulata = Cuarteron (capcious); Cuarteron + Mestiza (always quarreling) = Collote (strong, daring); Collote + Morisca = Albarazado (inclined to jokes and letdowns); Torna atrás (adult female) + Albarazado = Tente en el aire (ungrateful, evil).

to one of equality, but simultaneously moving from an organization based on ethnic state to one based on economic faculty (Florescano 2001, p. 229). The principle of the 'communal' is put under siege in favor of the 'individual'. The communal is contrasted even within the structure of the church. The suppression of the Inquisition is preceded by the expulsion of the Jesuites in 1767 and the suppression of thousands of confraternities, which had proliferated during the first evangelization effort at the cost of forms of syncretism with pagan credences adopting the fashions of a simple communal life in rural communities. The pillar of the colonial regime, the church, had offered the justification of the conquest and facilitated the inclusion of the natives in the social assemblage of the Vice-reign (Florescano 2001, p. 242). While the Neoclassical replaces the Baroque style, the intended restructuration of the church corresponds to a correspondent restructuration of the territory.

5.3.12 Forming the Image of the Patria Criolla

On the other hand, the Enlightenment brings the yearning for the rationalization of the administration of an "immense and poorly connected territory" (Florescano 2001, p. 244),[52] the govern of its many ethnic groups, and the exploitation of its resources. This leads to a vast collective effort of geographic exploration and mapping, started with a myriad of partial maps elaborated by *criollo* engineers until an overall chart of the homeland is put together by Carlos de Sigüenza y Góngora in 1748 and later improved by Francisco Díaz de León (Lombardo de Ruiz and Terán Trillo 1996, p. 33). Góngora's map will be the base for the delineation of the new administrative sub-divisions of *intendencias* and *provincias internas*. The elite of enlightened *criollos*, presiding the new cultural institutions of the colony and preparing for the struggle for independence, is the source that supplies Alexander von Humboldt's encyclopedic assemblage of the sociopolitical Atlas of the kingdom of New Spain, which will render the first overall image of the Spanish colony of worldwide diffusion.[53]

Intellectuals such as José Mutis, a Spanish-born botanist established in Bogotá, and Francisco de Caldas, a *criollo* naturalist from Popayan who will become a

---

[52]Translation by the author.

[53]Humboldt's work on Nueva España, published in Paris in 1811 is composed of three parts: a series of tables of data, titled *Tablas geográficas políticas del Reino de Nueva España que manifiestan su superficie, población, agricultura, fábricas, comercio, minas, rentas y fuerzas militares*; a series of twenty geographic maps and various diagrams, profiles and vistas, titled *Atlas géographique et physique du royaume de la Nouvelle Espagne*; and a critical essay synthesizing information from multiple fields, titled *Essai politique sur le royaume de la Nouvelle Espagne*.

martyr of the independence of Nueva Granada,[54] part of the same enlightened elite, had already supplied von Humboldt with all the available information during his adventurous exploration of the equinoctial regions. While the cultivated European public acclaims von Humboldt's presentations and books about the exotic continents of the Spanish colonies, his majestic inventory of the 'natural resources, developments of the cities, advancement of mining activities, circulation of goods, and progress of craftsmanship and industry' is saluted by the elite of Nueva España as the "irrefutable proof of the grandiose destiny reserved to the *patria criolla*"[55] (Florescano 2001, p. 246).

5.3.13 The Construction of the Estado Nacional

The fracture in the territorial regime, starting to surface with the Enlightened Bourbons, becomes conspicuous at the consummation of the independence in the 1820s, when the martyrized fathers of the *patria criolla* yield the way to the law-makers of the *estado nacional*. The idea of the *patria criolla* leads the *criollo* elites, in Mexico as in all the Hispanic colonies, to rise against the Spanish domination at the beginning of the nineteenth century seeking independence. The *estado nacional* is the institutional construction brought forth after the independence by the same *criollo* elites for about a century until the Mexican Revolution.

When Miguel Hidalgo adopts the ensign of the Virgen de Guadalupe to combine "ancient Mesoamerican resonances with Christian religion" in 1810, and José María Morelos attaches "the emblem of the foundation of Tenochtitlán" onto the insurgents' flag in 1812, the fathers of the *patria criolla* are summoning the idea of a 'mythical native nation'. But when Augustín de Iturbide proclaims the independence in 1821, the "historical project of the independence" converts itself into the project of "creating a modern nation and state" by using "a mix of ancestral myths, traditional patriotic drives, and religious symbols of identity"[56] (Florescano 2001, pp. 282–3).

5.3.14 Redistributing Ecclesiastic Properties

After the disastrous defeat in the US-Mexico war and the consequent 1848 cession of over a half of Mexico's territory to the United States of America, the persistent

[54]De Caldas was the first director of Bogotá's astronomical observer built in 1803 in the precincts of the botanical garden constituted by Mutis following the multiple botanical expeditions in the Nuevo Reino de Granada started in 1873.

[55]Translation by the author.

[56]Translation by the author.

attempt at a thorough restructuration of the remaining territory occurs along two major lines: the redistribution of ecclesial lands, including a half of the national territory in application of the Leyes de Reforma during an extended period from the late 1850s until the turn of the century; the violent occupation of lands cultivated by native groups for their own subsistence according to communal regimes and officially unclaimed, operated by the *criollo* regional elites.

The civil war that breaks out from the permanent conflict between the federalist liberals and the centralist conservatives will see the liberals eventually prevail. In a period of uncertain fate between 1854 and 1873, the Leyes de Reforma are promulgated by the liberal government from the capital or from exile, a series of radical laws inspired by the ideas of the French Revolution.[57] Initiated and institutionalized by Sebastián Lerdo and principally actuated by Benito Juárez, the laws concern the separation of state and church, including the suppression of all religious orders and the nationalization of the ecclesial property that extended over a half of the country. While conceived for an equitable redistribution of the ecclesial properties, the implementation of the Reforma will result in auctions to the best offeror favoring the accumulation of latifundia in the hands of few, largely foreign, holders.[58] The Leyes de Reforma are substantially confirmed by the conservative governments of the Second Empire, supposedly ally of the church[59]

5.3.15 *Against the Communal Ways of Life*

The idea of dismantling the 'ecclesiastical' probably lies on the conviction that the pillar of the colonial regime is not anymore necessary to the national state. At the demise of the colonial, even the 'real native' is replaced by the mythical native

[57]Rosa María Martínez writes that when the cycle of the Leyes de Reforma are completed in 1873 with the annexation into the Constitution, five short articles sanctioned: independence of the State from the Church; secularization of marriage and civil state; prohibition for religious institutions to possess estates other than churches and residences for priests and bishops; replacement of the civil for the religious oaths (Martínez 2008, p. 142). To this, we can add the secularization of cemeteries, suppression of monasteries, exclaustration of nuns and institution of the calendar of laic public holidays.

[58]Martínez asserts that the Ley de Nacionalización de Bienes del Clero y su Reglamento deliberated by the Juárez government in 1859 happened to be much more effective than any previous laws, almost resulting in a general confiscation. The law's preamble states how the practical purpose is to pass all the possessions of the Church into private ownership through immediate auctions and destine the profits to cover civil war expenditures. This leads to the consideration that the spirit of the Ley de Nacionalización is not that of increasing the federal patrimony, but rather to render the properties available to entrepreneurial initiative, while rapidly cashing to cover a substantial national debt (Martínez 2008, pp. 140–1).

[59]According to Martínez, the Second Empire (April 1846 to July 1867) does nothing but consolidating Juárez' nationalization model, which will then be confirmed once again with the restauration of the Republic. Therefore, the State had been confiscating and selling religious properties without interruption. (Martínez 2008, pp. 141–142).

nation of the past. The redistribution of ecclesial lands is then tackled in parallel with a violent aggression onto the lands of the natives and their communal ways of life. "At the fall of the colonial State and with the disappearance of the Leyes de Indias, which protected the peasant republics", the communal lands of the indigenous remained unprotected. "By declaring equal citizens all the inhabitants of the republic", the Leyes de Reforma follow the same trans-historical instances stripping the ethnic groups of the rights to their communal living forms. In its complex, the process of edification of the *Estado Nacional* can be said to move in the direction of "appropriating the land of the indigenous, destructing the institutions that cohesed their ethnic identities and combating the traditions, the culture, and the values of the indigenous" (Florescano 2001, pp. 429–31).[60]

As for the expropriations that change the urban structure of Paris in the same years, these territorial transformations embody trans-historical tendencies implemented across time by different and even opposed groups of power: in the case of Mexico, the aspiration to the forging of a modern national state that we have mentioned.

5.3.16 Lost Idylls of the Yucatan Peninsula

In the southern peninsula of Yucatan, during the *colonia* subsistence agriculture had been practiced by rural communities on communal or unclaimed lands in parallel to the obliged service for the commercial economy of *encomiendas* and *estancias*. This activity suddenly drops when in 1825 a law facilitates the acquisition of unclaimed lands by part of large and medium land owners, in 1844 new taxes are charged on the peasants, and in 1847 new laws reduce the extension of the communal lands by imposing the erection of property boundaries. The new laws are meant to favor the formations of larger *haciendas* for the production of sugar and henequen. The indigenous unrest that shakes Yucatan in the second half of the nineteenth century has various motives, among which the drastic drop of the quality of life of the communities. In this scheme, the bloody Guerra de Castas of Yucatan cannot be simply explained on the grounds of racial resentment. The ethnic violence has to be rather framed within the conflicts among opposite *criollo* elites for the supremacy on the territory, in which the indigenous are enlisted in one or the other faction with vane promises (Florescano 2001, pp. 425–9).

The attacks of the indigenous groups against the whites are then a defensive response to the attempt at the violent occupation of the ancestral lands, the obliteration of the ancient communal agricultural costumes and ways of tending the land, and finally the thorough alteration of the territorial structure.

[60]Translated by the author.

5.3.17 *Bucolic Communes Alongside the Yaqui River*

The violent struggle of the Porfiriato against the Yaqui people in the northern Mexican state of Sonora, at the geographic antipodes of Yucatan, is another blatant attack of the modern *estado nacional* onto the communal life of indigenous cultures. As for the case of Yucatan, however, the material object of the attack is the structure of the land, in which the communal regime is embedded.

The Yaqui people form an indissoluble unity with their lands alongside the Yaqui River. They conduct a productive communal life settled in eight towns governed by collective institutions. In 1875, in the face of the advancing *estado nacional*, they claim independence to their homeland under the charismatic leader Cajeme. The fertile fluvial lands of the Yaquis will soon attract the attention of Mexican and foreign investors. The army of the Porfiriato smashes the Yaqui utopia in 1887 and Cajeme is killed. Rather than being subjected to a capitalistic economy and the consequent territorial restructuration, the Yaquis abandon their lands on the banks of the river and disperse in the mountains. Lured into coming back as workforce with void promises ten years later, they find their communal homeland subdivided into individual plots, crossed by artificial canals and modern infrastructures, and occupied by the national army. To their demands for the liberation of their indivisible land, the Porfiriato responds with a brutal repression and the mass deportation of the Yaqui people to serve as enslaved workforce in the plantations of Oaxaca (Florescano 2001, pp. 407–13).

5.4 The Body of the Metropolis

5.4.1 *The Leyes de Reforma Resemble an Elegant Boulevard in French Style*

Should the practical consequences of the Leyes de Reforma on land patterns be evaluated through the microhistories of the diverse territories of the country,[61] their univocal urban embodiments can be found aligned along the Paseo de la Reforma in the capital. The implementation of the *paseo* can be summarized in three main phases: the first is relative to its conception, tracing, and foundation; the second to its implementation as an urban infrastructure in the service of the new urbanites, equivalent to the analogous programs of metropolitan improvement of European cities; the third to its monumentalization and relayering with the symbols of the

[61]At odds with the case of the Yaqui lands, in *The Reform in Oaxaca 1856–76*, Charles Berry concludes that the liberal revolution in Oaxaca is, by and large, a movement of the middle-class, urban and elitist, which did not show any interest in the effects on the life of the peasants and natives. Its effect on the social and territorial structure is only noticeable in the areas closest to the capital of Oaxaca and its institutions (Berry 1981).

identification of the Reforma with the construction of the modern *estado nacional*. The three phases are brought forth by political regimes of opposed orientation, respectively the Second Empire, the liberal government of the República, and the conservative government of the Porfirian dictatorship, imprinting trans-historical aspirations onto places and spaces of the Mexican capital.

As part of a French-inspired plan for urban axes irradiating from the *zócalo*, the monumental boulevard is conceived at the beginning of the Second Empire to connect the city center and the residency of the Habsburg emperor in the castle of Chapultepec. By 1867, a suburban way reserved to exclusive use of the monarch is roughly laid down over lands reclaimed from the Texcoco lake and occupied by crops and cattle farming and crossed by canals.

After the fall of Maximilian's regime, the way is open to the public and by 1870 its first half, from the central Alameda to the Glorieta de la Palma, is turned into a boulevard with tree-lined sidewalks. In the following two years, under the restored republican government of Lerdo, the plan of an elegant boulevard reaching the forest of Chapultepec is finally implemented with lavish amenities including broad sidewalks equipped with benches and abundant vegetation on both sides and monumental *glorietas*. The development of elegant residential districts of French taste, the Colonia Americana on the south side and the Colonia Cuauhtémoc on the north side of the boulevard, are promoted. In 1872, the boulevard is eventually renamed Paseo de la Reforma by president Lerdo, sanctioning a correspondence between the law-making action and the reinvention of the monumental topography of the capital, to secure the space for democracy that Arendt elaborates on.

However, it will be the conservative president Porfirio Díaz, and *de facto* dictator of the country for about 30 years, to run a vast sculptural program for the decoration and monumental equipment of the *paseo*. Among numerous pieces, the installation of the monumental statues of Cuauhtémoc and Columbus on two *glorietas* in close proximity, is particularly representative of the distinct and coexisting drives at the basis of the construction of the *estado nacional*. A modern entity that simultaneously pretend to rise from the mythical pre-Hispanic nation and the European lineage of the *criollo* and *mestizo* elite. The consolidation in consistent French style of the new residential areas on the two sides of the urban axis will end up forming the *colonias* of La Condesa, Hipódromo, and Roma, the most elegant neighborhoods in the historical city today. The completion of the Paseo de la Reforma is ideally established with the installation of the Puerta de los Leones, the cast iron gates to the forest of Chapultepec in 1892.

As for Rue de Rivoli in Paris, the Paseo de la Reforma in Mexico City is the material precipitation of trans-historical aspirations across changing regimes. The monumental transcription of the modern national state's edification into the structure of the Mexican capital, echoes, in grand symbolic forms, the process of cultural miscegenation imprinted into the productive organization of land of the Mexican territory at the geographic scale.

5.4.2 The Venerated Forest of Chapultepec

Based on the histories of Juan de Torquemada and Diego Durán, Miguel Angel Fernández tells us that, long before becoming the terminal of a metropolitan axis, Chapultepec had been the mountain (*petl*) of the grasshopper (*chapul*), sitting on a cave of crystal waters and surrounded by a forest of tall cypresses of the ancient accounts. The cave, the forest, and the mountain had been venerated as a living deity since the first inhabitants settled on the shore of the lake. They had been caring to the place by pruning their divine forest and cleaning the cave. Fresh springs and fertile lands had nourished the crops and flowers of an idyllic place that could donate eternal life. At the same time, it had been also the scenery of imperial burials and human sacrifices (Fernández et al. 1988, pp. 15–9).

Chapultepec is then the ancestral *altépetl*, the cosmic axis aligning the depths of the lake, the forested and fertile land, and the rocks of the relief pointing towards the sky. By placing the monumental portals to the sacred forest of Chapultepec, modern Mexico expresses the unconscious aspiration to reconnect the *criollo* edification of its *estado nacional*, to the ancestral structure of a mythical pre-Hispanic nation.

5.5 Chapter's Conclusions

5.5.1 Correspondences and Divergences Between Patterns of Space-Forming

Divergences prevail over correspondences in the space-forming patterns of Mexico. The offended and wasted hinterlands of the Republic do not resonate in the creole livery of the *zócalo* of the Marquesado del Valle, the urban ensemble of the city of Cortés, nor in the monumental transcription of the nation along the metropolitan axes of the capital and their grand narrative equipment.

In their moving book about the Mexican plaza (Wagner et al. 2013), or more precisely about the nature and formation of urban open space in Mexican urban settings, Wagner, Box, and Morehead reconstruct the transformative sequence that the Mesoamerican centers are subjected after the fall of the Aztec empire at hand of the waves of mendicant friars that scatter into the hinterlands, nestle in the most remote location and devote their existence to the parallel implementation of an unprecedented campaign of proselytism and the "largest campaign of town founding the world would ever experience" (Wagner et al. 2013, p. 49). The first act of what we could call the reterritorialization of Nueva España is the erection of a Christian cross onto the summit of the Mesoamerican pyramid, new *axis mundi* to replace the Tree of Life. Then a rudimentary altar sheltered by a thatched-roof that will be identified with the entrance to the primordial cave. A wall will be drawn to enclose the sacred patio of the *atrio*, and four chapels set at the corners to form a *quincunx* while serving the ceremonial stops of the counterclockwise procession of the via *sacra*. The *atrio* is probably the epitome of the religious and spatial miscegenation of the Mesoamerican and Spanish cultures: the ancient peregrination is transcribed into the via *sacra*, the corner chapels of the Christian cult mark the Mesoamerican X from intercardinal points across the sacred ground, as the primordial ocean, and all revolves around the focus of the cross, new *axis mundi*, replacing the venerable *nopal* laden with prickly pears. The direct visual relation with the natural world and its deities, the mountains, the forests, the waters, the sky is unimpeded in the open spatial formation of the *atrio*. Man's footpaths across it also remain legible. The convent will be built starting with the cloister, the most private open space of the town, and terminating with the nave of the church last, since the open space of the *atrio* perfectly suits the Mesoamerican customs of open air religious and civic celebrations in the meantime. Then comes the secular plaza next to the religious complex, and also the market place, the *plaza de armas* (parade ground), separated of fused in one urban space (Wagner et al. 2013, pp. 47–48). Last term of the sequence is the *alameda*, the provision of natural services for the urbanite, Hispanic version of the metropolitan public park of the bourgeois city.[62] From that nucleus the colonial city would expand a square grid of urban blocks that subdivide the communal lands of the natives often erasing the layered memory, sometimes adapting to preexisting elements and formations. The Relaciones Geográficas of the sixteenth and seventeenth centuries, through which the religious and civic functionaries of the colony are required to illustrate the state of the new territories, are accompanied by diagrammatic pictorial boards mixing planar and tridimensional illustrations showing the rigid urban grid over-imposed onto the natural topographies and fluid footpaths following its curves.

On one hand, the urban assemblage of the *zócalo* of Oaxaca and the metropolitan accumulations along the Paseo de la Reforma culminating in Chapultepec's forest and mound, are the spatial manifestations of the most sublime achievements of the ethno-cultural miscegenation that has molded the centers of Mexico through waves of deterritorialization and reterritorializations. On the other hand, large hinterlands of the country represent the aborted evidence of a historical sequence of collective convulsions that have tragically envisaged the progressive reform Mexico's geography, while seizing its land for the henequen plantations in the Yucatan peninsula or ripping it in pieces along the Yaqui River, and rapaciously marauding its immense natural variety.

[62]About the introduction of public parks (*Volsksgarten*) as supplies of natural services for the metropolitanites of modernizing cities since the beginning of the eighteenth century, see Chap. 7 of this book.

5.5.2 *A Doleful Stream of Trans-Historical Aspirations and the Geography of Miscegenated Continents*

What converts the complex Mesoamerican ethno-geography into the present post-colonial territory is, by and large, a doleful stream of simultaneous histories of brutal hybridizations culminating in the 'sad waltz'[63] of the Mexican Revolution and its massacres. But the superb image of the *patria criolla* summoned by the Humboldtian enterprises still rises from it, upholding the grandiose destiny of the New World. It is in fact von Humboldt's words that Iñaki Ábalos and Juan Herreros resort to, to describe the magnificent promise disclosed by the New World: the geography of 'miscegenated' continents will only be depicted by the landscape painting "when the great artists cross the frontiers of the Mediterranean [...], when they are able to embrace the immense variety of nature." (Ábalos and Herreros 2003, p. 52) In that miscegenation, the conglomerate of nature and artifice confused, will then illustrate the "dark zones of the atlas of the picturesque [and an] immediate, unified sense of beauty" (Ábalos and Herreros 2003, p. 57).

The creole self-agency of Baker, doing splits in the air veiled by nothing but a pink ostrich feather, shimmying behind a shiny string of artificial bananas, hiding war secrets in her music sheets, receiving the Rosette de la Resistance, and speaking in her Free French Army uniform at the March on Washington, embodies the culture-forming struggle, parallel to the region-forming struggle, kindled by the violent action of colonization. Thereby, the colonized re-determines a self-agency, re-molds an own body and a novel behavior, and conquers the colonizer in turn. We are then allowed to conclude that Baker's creole self-agency is the equivalent of the spatial complex of the *zócalo* in Oaxaca and the axial composition of the Avenida de la Reforma in Mexico City, that is the summits of the physical stratifications of culture- and region-forming processes of miscegenation. On the other hand, are the Mexican hinterlands, dominated by unresolved exploitative behaviors of man over geography, preventing the solution of the struggle in a 'unified sense of beauty' and obstructing the tracing of an atlas of the picturesque.

References

Ábalos I, Herreros J (2003) Journey through the Picturesque (a notebook). In: Mostafavi M, Najle C (eds) Landscape Urbanism: a manual for the machinic landscape. Architectural Association, London

Abulafia D (2011) The great sea: a human history of the mediterranean. Oxford University Press, London

[63]Quoting Miguel Ángel Berumen, curator of the photographic traveling exhibition *La Revolución Mexicana: Un Valse Triste*, 2015 [in English: The Mexican Revolution: a Sad Waltz].

Barnwell A (1997) Like the Gypsy's daughter: or beyond the potency of Jose-phine Baker's eroticism. In: Skipwith J (ed) Rhapsodies in black: art of the Harlem Renaissance. Hayward Gallery/Institute of International Visual Arts/University of California Press, Los Angeles
Bastide JFD (1994) The little house: an architectural seduction (trans: El-Khoury R). Princeton Architectural Press, New York
Bernal R (1969) El complot mongol. Joaquín Mortiz, Ciudad de México
Bernal R (2012) El Gran Océano. Fondo de Cultura Económica, Ciudad de México
Bernal R (2015a) Antología policíaca. Fondo de Cultura Económica, Ciudad de México
Bernal R (2015b) Mestizaje y criollismo en la literature de la Nueva España del siglo XVI. Fondo de Cultura Económica, Ciudad de México
Berry C (1981) The reform in Oaxaca 1856–76. University of Nebraska Press
Birksted J (2009) Le Corbusier and the Occult. MIT Press, Cambridge
Broughton S (Escritor), Phillips S (Dirección) (2006) Josephine Baker [Película]
Burns K (1997) A house for Josephine Baker. In: Nalbantoğlu GB, Wong CT (eds) Postcolonial spaces. Princeton Architectural Press, New York, pp 53–72
Benevolo L, Romano S (eds) (1998) La cittá europea fuori d'Europa. Libri Scheiwiller, Milano
Conforti C (2017) La casa di Anna e altre belle storie. Rassegna di Architettura e Urbanistica, vol 52 (152, Vittorio De Feo)
Dalton K, Gates H Jr (1998) Josephine Baker and Paul Colin: African Ameri-can Dance Seen through Parisian Eyes'. Critical Inquiry 24(4):903–934
De Feo V (2010) Tre racconti di architettura (trans: 1st edn. 1997, Belardi PB). Edizioni Libria, Melfi
De Maria, Campos Castelló A (2015) Prólogo a la edición de FCE. In: Bernal R (ed) Mestizaje y criollismo en la literature de la Nueva España del siglo XVI. Fondo de Cultura Económica, Ciudad de México, pp 15–23
Deleuze G, Guattari F (1987) A thousand plateaus: capitalism and schizophrenia (Trans. Massumi B). The University of Minnesota Press, Minneapolis
El-Dahdah F (1995) The Josephine Baker house: for Loos's pleasure. Assemblage 26
El-Khoury R (1994) Introduction. In: Bastide JFD (ed) The little house: an architectural seduction. Princeton Architectural Press, New York
Fernández MA et al (1988) Chapultepec: Historia y Presencia. Smurfit Cartón y Papel, Ciudad de México
Flanner J (November de 1925) Paris letter November 14, 1925. The New Yorker
Florescano E (2001) Etnía, Estado y Nación, 1st edn. Taurus, Madrid
Gibson C (1995) The transformation of the Indian community in New Spain, 1500–1800. Cahiers d'histoire mondiale II:581–607
Hale C (1968) Mexican liberalism in the age of Mora, 1821–1853. Yale University Press, New Haven
Hale C (1989) The transformation of liberalism in late 19th-century Mexico. Princeton University Press, New York
Henderson M (2003) Josephine Baker and La Revue Negre: from ethnography to performance. Text Perform Quart 23(2)
Hight C (2003) Portraying the Urban landscape: landscape in architectural criticism and theory, 1960–present. In: Mostafavi M, Najle C (eds) Landscape Urbanism: a manual for the mechanic landscape. Architectural Association
Lafaye J (1976) Quetzalcóatl and Guadalupe. The formation of Mexican National Consciousness, 1531–1813 (trans: French: Gallimard 1974 edn., Keen B). The University of Chicago Press, Chicago
León-Portilla M (2008 [1959]) Visión de los vencidos: Relaciones indígenas de la conquista (trans: 1st edn., UNAM 1959, León-Portilla MED). Ediciones Digitales UNAM
Lockhart J (1992) The Nahuas after the conquest. A social and cultural history of the Indians of Central Mexico, sixteenth through eighteenth centuries. Stanford University Press, Palo Alto
Lombardo de Ruiz S, Terán Trillo Y (1996) Atlas histórico de la Ciudad de México, vol II (de la Torre M ed). Smurfit Cartón y Papel, Ciudad de México

Martínez RM (2008) Juárez, su obra, su tiempo, y su mundo jurídico. Las Leyes de Reforma. Anuario Mexicano de Historia del Derecho 20:129–147
Mohar M (1999) El Mapa Quinatzin. De valientes guerreros chichimecas a sabios y poderosos gobernantes. tesis de doctorado en Ciencias Sociales y Políticas. UNAM, tutor Carmen Viqueira, Ciudad de México
Mostafavi M, Najle C (ed) (2003) Landscape urbanism: a manual for the mechanic landscape. Architectural Association, London
Romano S (1998) Dopo il trattato di Tordesillas: l'emisfero spagnolo (introduction to Part III). In: Benevolo L, Romano S (ed) La città europea fuori d'Europa. Libri Scheiwiller, Milano
Romano S (1998) Preface. In: Benevolo L, Romano S (ed) La città europea fuori d'Europa. Libri Scheiwiller, Milano
Roueff O (2006) Politiques d'une "culture nègre": La Revue Nègre (1925) comme événement public. Anthropologie et Sociétés 30(2):65–85
Sahagún BD, informantes. (compiled 1577). Capítulo 11 De como los españoles llegaron a tlaxcalla que entonces se llamaba texcalla. En Códice Florentino or Historial general de las cosas de Nueva España. World Digital Library, Biblioteca Laurenziana di Firenze, pp spreads 423–424
Sowinska A (2005–06). Dialectics of the banana skirt: the ambiguities of Josephine Baker's self-representation. In: Michigan feminist studies 19 (Bodies: Physical and Abstract)
Sweeney C (2004) From Fetish to subject: race, modernism, and primitivism, 1919–1935. Präger, Westport, CT
Tafuri M (1972) Design and technological Utopias. In: Ambasz E (ed) Italy: the new domestic landscape. MOMA, New York
Vidler A (1994) Preface. In: Bastide JFD (ed) The little house: an architectural seduction. Princeton Architectural Press, New York
Wagner L, Box H, Morehead SK (2013) Ancient origins of the Mexican plaza. From primordial sea to public space. University of Texas Press, Austin

Legal Notice:

Fig. 5.1: Lucien Waléry (https://commons.wikimedia.org/wiki/File:Baker_Banana.jpg), "Baker Banana", marked as public domain, more details on Wikimedia Commons: https://commons.wikimedia.org/wiki/Template:PD-1996

Chapter 6
Gardens Grown Wild: In-Between Topography and Its Mythology

Abstract This chapter uses the construct of the garden to explore the correspondences between physical topography and its mythology, reality and its transcription, in contained and uncontained landscapes. The text analyzes the articulate assemblage of the ancient landscape in the Bay of Naples stretching from the urban spaces and art apparatuses of the city of Pompeii, across the semi-wild terrains of the shore beyond its walls, and out into the atmosphere above the waters. Elaborating further, the text also articulates the reciprocal reflection of man-made and natural wefts and the confluence of history and autobiography in an atmospheric time proper of the Pompeian landscape. The text, then, discusses two alternative ideas of garden, the nature garden and the people's garden, corresponding to a shift in ethos and two opposite ways of regarding nature in more general terms. The latter corresponds to man regarding nature as a provision of natural services, the former to considering man as an observer immersed in the natural ecosystem. The text, then, interprets the historical transition converting the hunting parks into metropolitan parks in the European capitals, going from game hunting to service providing, as a consolidation of the feudal structure of thought into the modernity. In conclusion, ethical alternatives to the unsustainable exploitation of natural services can replace the understanding of the world's relational dynamics for the careless action of consumption of its resources.

Keywords Nature garden · People's garden · Pompeii · Parietal painting
Painted garden · Meteorological time · Hunting garden · Subtle hunts

6.1 Introduction

This chapter renders a vision of the contemporary continuum as a broken mirror reflects the image of the world: each shard of the mirror reflects a complete image of the universe, while it is also part of an overall mosaic. Fragment and totality coexist. In the mirror of the landscape, the garden plays the role of the shard. It plots the correspondence between fragment and totality (Fig. 6.1).

R. Pasini, *Landscape Paradigms and Post-urban Spaces*, The Urban Book Series,
https://doi.org/10.1007/978-3-319-77887-7_6

Fig. 6.1 Frontispiece of *Social Life in the Insect World*, London 1912, English anthology of Jean-Henri Fabre's *Souvenirs Entomologiques. Étude sur l'instinct et les moeurs des insects*, a series of books published in the period 1879–1909. The illustration summarizes observations about various 'social' attitudes of the Mantis (*image source* social Life in the Insect World, London, T. F. Unwin 1912, Not in Copyright, PD)

In his posthumously fortunate essay *Of Other Spaces: Utopias and Heterotopias* (Foucault 1984),[1] Foucault refers to the garden as the most ancient example of heterotopia that is a heterotopia where several juxtaposed sites coexist in the same real space.[2] According to Foucault, if the traditional Persian carpet is a symbolic representation of a garden, the garden is also a carpet in which the universe achieves its symbolic perfection. Therefore, the tiniest parcel of the world, the garden, coincides with the totality of the universe. The garden carpet, third articulation of the taxonomy of heterotopias, revolves about the indissoluble coexistence of the real site and its symbolic transcription. The same construct can be easily applied within or beyond the garden fence.

The remapping of a landscape assemblage through the garden paradigm shows the cracks in reality that lead to the cavities between the site and its semantic narrative.

6.2 Writing the Universe Within and Beyond the Garden Fence

6.2.1 Phenomenon and Text, Gardens of Pelion

The short description of the communal garden at Agios Ioannis, on the coast of the Pelion region, and its quiet dynamics, lends the title to the notebook of quotidian observations *Il giardino del Pelio* by Pietro Bellasi (Bellasi 1987, pp. 17–9). The Pelion observations draw a phenomenological representation of the interstices generated by the coexistence of reality and its writing. Bellasi's notebook records the 'intellectual art' of an uninterrupted observation of the quotidian, as subjectively and affectively vibrant as to uncover the forms that write our incessant day-by-day narratives (Bellasi 1987, pp. 13–4). Bellasi describes a peculiar garden, as a gash opening a rural oasis in the urban fabric of a Greek small town. Contained by the rears of diverse houses and hotels, the site shares the characters of both the wild and cultivated garden, where dismissed urban junk accumulates in a novel rural life and ushers the return of Mediterranean *maquis*. There, a kid runs following a trail barely marked through sagging fences, casual fruit trees that compose woods instead of orchards, clumps of shrubs and herbs. Clothed by the foliage of a group of pear and apple trees, the kid's trajectory does not reappear where the path suggests, but rather dashes for a moment much further away, before definitively vanishing from sight. The acoustic perception of the observer is also deceived by the same

[1]The text was published in the magazine Architecture, Mouvement, Continuité in 1984, but presented as a conference with the title of *Des espaces autres* at the Centre d'Études architecturales in Paris in 1967. The reference here is to the English translation by J. Miskowiec.

[2]Catharine Edwards and Greg Woolf's idea of Cosmopolis also falls into such a type of heterotopian space (Edwards and Woolf 2003).

obstacles. By nesting wormholes in the perception of the scene and generating 'labyrinths' in the transcription of the quotidian, such little everyday mysteries illuminate the discrepancies of phenomenon and text, or artifact and plan, as well as the ways of their coexistence.

6.2.2 Beyond the Garden Fence

By letting our gaze slip beyond the fence, we extend the garden construct over uncontained landscape assemblages and their narratives. A thin fine book by Annamaria Ciarallo has been sitting on my desk for years. *Il giardino pompeiano. Le piante, l'orto, i segreti della cucina* (Ciarallo 2002) could be described as a sentimental exploration of Roman gardening seen through the lens of Pompeian archeological heritage. The book is structured in three main parts.

First, a critical excursion through Lucius Columella's treaty on agriculture (*De re rustica*, first century BC) focuses on Books X and XII dedicated to horticulture and food preparation. The characteristics of a good suburban *hortus* are described along with an almanac of the seasonal activities, methods for a long conservation of the produce, and its preparation in homemade recipes.[3] A second part reviews a set of physical characteristics of the Roman suburban garden, which, collected through a series of Pompeian cases, results in the determination of a type. The final part presents a catalog of ancient Pompeian flora, reconstructed from parietal paintings and paleobotanic analyses, compared with the present-day flora of the archeological area. The archeological area at Pompeii has, in fact, retained a remarkable biodiversity compared with its environs affected by the agricultural use of chemicals.

6.2.3 Insula Occidentalis 42

The book is illustrated by a splendid apparatus of photographs that reproduce the vivid garden scenes frescoed on the walls of Pompeian villas. The series of stunning images from the open *triclinium* and *oecus* of the House of the Golden Bracelet plunges the entire book in an oneiric dimension.

In another book, 'Insula Occidentalis 42' (Ciardiello 2006, pp. 69–256), Rosaria Ciardiello documents with philological detail the archeological analysis of that house. At the end of the second century BC, the geopolitical stability of Campania allowed for the development of an *insula occidentalis*, a residential block, on the western hillsides of Pompeii, overlooking the Tyrrhenian Sea. The houses of the

[3]As Ciarallo notes, in the *De re rustica*, a literary tradition of agricultural almanacs that remounts back to Hesiod's eighth-century BC *Works and Days* is combined with a practical manual for Victorian ladies. On the other hand, Virgil, Columella's contemporary, keeps his *Georgics* on a lyric key.

insula in fact could be installed atop the by-then-unnecessary fortified walls used as substructures. We could say that the convergence of nature, topography, monument, and architecture was starting to accumulate upon that location a unique series of spatial characteristics generating a *locus*. The subsequent phases of extension and refinement of the residential structures until the eruption of 79 consolidated an extraordinary sedimentation of a primordial natural platform and urban artifacts with unparalleled spatial works of art, including sophisticated gardens, statuary, and parietal paintings.

6.2.4 Open Your Summer Triclinium and Oecus

The House of the Golden Bracelet in the *insula occidentalis* constitutes an extraordinary assemblage of architecture and art within the bearing frame of that natural/urban *locus*. Located at the very end of the Via Thermarum, the architectural *machina* of the house led the visitor through three levels, from the buzzy urban environment above to a sphere *autre* below, where one was projected into an idyll with nature. The very architectural interface between urbanity and nature, or anthropogenic and geogenic realms, precipitated in a condensed space at the bottom level. There, the façade exhibited the concavities of a *triclinium* and a summertime *oecus*, open onto a terraced garden overlooking the shore. The barrel vaults of both rooms still establish an uninterrupted transition that projects the beholder's perceived space from the interior of the building out onto the platform of the terrace. From there, the spatial perception follows the sloping hillside covered with Mediterranean *maquis* down to the shore, to then be cast into the airy and watery atmosphere above the Bay of Naples.

6.2.5 Water Whirls and Lava Foam Spring Out of a Vitreous Recess in the City Wall

The *triclinium* is the very center of the overall assemblage and the privileged vantage point of the perspectival space that unfolds. There, the architecture of the house merges with the environmental system. The back wall of the room opens in a curved niche occupied by a *nymphaeum*. In its center, a stepped fountain in the shape of a cascaded spring is recessed deep into the substructures. From it, water used to stream out into a rectangular basin. The niche was covered with vitreous tiles arranged in polychrome mosaics representing plants, while the cascade was clad in lava foam mimicking a natural cave. The cisterns alimenting the water features were built in the interstice between two septa of the unused defensive walls, so that the assemblage of architectural, urban, and natural elements orchestated in the grotto was indissoluble.

6.2.6 'Parietal-Garden-Paintings/Cultivated-Vines/ Spontaneous-Maquis'

The fixed beds of the *triclinium* are still intact, U-shaped around the rectangular basin and separated from the outside by another fountain pool with metal nozzles that used to sprinkle water. The interior walls were covered with garden paintings of a hyperrealistic vividness. Large *lacerti* of the paintings have survived mounted on panels in the restoration laboratories of the superintendence of Pompeii. Luxuriant vegetation, flowers, birds, and various exotic figures, such as hermae, *pinakes*, *oscillae,* cupids, Egyptian and Greek sphinxes, were framed by the slender columns and light structures of an illusory wooden canopy. The painted canopy extended into a real pergola projecting out onto the terrace garden to provide shade. The garden was flanked north by a high wall covered with cultivated vines that rambled over the pergola, so that the real foliage mixed up with the painted vegetation in the interior. Therefore, the flora and fauna depicted in the parietal paintings resonated in the 'second nature' of the cultivated plants of the terrace garden and, beyond it, in the 'first nature' of the spontaneous *maquis* covering the western hillsides of Pompeii.

6.2.7 Longitudinal Trajectory Through Vertical Facialities

The larger assemblage of 'lava/urban-walls/water-cisterns/cascade/vitreous-*nymphaeum*/parietal-painting-garden/hanging-garden/*maquis*-covered-hillside/Tyrrhenian-shore/open-sea/mist-of-airy-and-watery-atmosphere/nocturnal-asters' organizes the totality of a landscape around the synesthetic perception of the civilized beholder lying on the *triclinium.* The Pompeian landscape kindled by the House of the Golden Bracelet is a vertical accumulation of layers allowing for a longitudinal projection of the beholder. A longitudinal trajectory of reconnaissance through the organization of 'facialities' (birds, flowers, patricians, hermae, slaves, Egyptian gods) spans the Jungian series nature-culture-nature ('wild-*maquis*/cultivated-garden/garden-paintings') that organizes the collective imaginary.[4]

6.2.8 Of Meteors Between Lava and Asters

Poising in-between lava and asters, the Pompeian landscape is suspended in the atmospheric time of water moving in waves, whirling air, weather, turmoil of multiple particles, clouds forming. It is the 'time of meteorology' that Deleuze

[4]About the nature-culture-nature series, see subchapter 'Analogies by resemblance and proportion' in Chap. 8 of this book.

Guattari comment upon: between "the extreme slownesses and vertiginous speeds of geology and astronomy, Michel Tournier places meteorology, where meteors live at our pace" (Deleuze and Guattari 1987, p. 261). Tournier closes *Gemini*, his novel of the man/natural correspondences, by returning to the refrain of the description of the transient evolution of the 'meteors'.

A depression moving from Newfoundland to the Baltic had opened Tournier's book. 'Masses of warm, moist oceanic air are pushed into the corridor of the English Channel and a gust of wind from west-southwest turns over, among other things, eight pages of Aristotle's *Meteorologica* that the author is reading on the beach at Saint-Jacut.'[5] (Tournier 1997, p. 9).

Episodes from the life of two anticyclones close the book. 'One situated off northern France, the other off the southwest of England, Arctic fortresses—glacis of still, cold air—stoutly withstanding the onslaught of the Atlantic currents moving into fill a deep depression. The one off Cornwall surrendering to warm air, breaking and teetering on the edge of the gulf of low pressure. The Flemish one remains impregnable, "strong in a high-pressure area of 1021 millibars", directing a calm, clear flow of dry, freezing wind towards the author. Yet the covering of snow on the fields is thinning due to the bright, keen sun causing the snow to evaporate without thawing. A phenomenon known as sublimation'[6] (Tournier 1997, p. 452).

In Tournier, the unfolding of man's life is like the formation and dissolution of nimbuses: "A cloud forms in the sky like an image in my brain, the wind blows like I breathe, a rainbow spans the horizon for as long as my heart needs to reconcile itself to life, the summer passes like vacation drifts by" (Tournier 1997, p. 450). The same meteorological dimension has governed the landscape assemblage of the Bay of Naples.

6.2.9 Of Forms and Language

The conjunction of collective history and autobiographical memories occurs in the meteorological assemblage, where the scientific observations of the object correspond to the subject's aesthetic contemplation—science and autobiography no longer at odds.

I bought the little book with the parietal garden paintings during a summertime Neapolitan trip. I was lodging in an eighteenth-century villa uphill of the historic Rione Sanità in Naples. Every morning I walked downhill the tiny streets from Capodimonte to Plebiscito. Moiariello, Morisani, Riccia, Miracoli, Cristallini, Antesaecula, Sanità, Vergini: architectural gems popping out of the miserable ensemble stratified over the hollow topography of the mound, beauty spilling from the portals of churches in waves of glittering *ex-voto*, velvets, and molded stones.

---

[5]Paraphrased by the author from *Gemini*

[6]Paraphrased by the author.

Along the descent to the sea, I used to recall Wilhelm Tischbein's letters recorded by Goethe in the *Italian Journey*. A Moorish lady had been once kept captive on a seized boat in that same bay before my eyes. Crowds drifted on dinghies to see her, some affording a prize.[7] Along with this urban narration, another one is set outside of the city amid the surfacing ruins of Baia Domitia. A goatherd had come onto the seashore at the mouth of the Garigliano River. The goats dove into the water. There arrived a swineherd too. While the herds sought refreshment in the waves, the shepherds lay in the shadow playing music. Eventually, a handsome young man came forward on horseback and submerged himself so deeply into the sea that even the horse seemed to disappear with him. Only the horse's head and the youth's shoulders were out of the water. (Tischbein 1993, pp. 446–7)

Inside or outside of the city, the 'beautiful spectacles' are about the spell of interacting forms, that is, constructions, topography, objects, and bodies of men and animals, which shape the language of the observer. "From Italy, rich in forms, I was thrust back into formless Germany—writes Goethe several years after his return—and obliged to exchange a bright sky for a somber one. [M]y friends brought me to the brink of despair. My rapture over objects that were far away [...] seemed to offend them; [...] no one understood my language"[8] (Goethe 1993, pp. 572, 542 note 1). Goethe's passage points out the correspondence between the forms of the landscape and the language that transcribes it.

6.2.10 Meteorological Mysteries

Of the visit in Pompeii, I remember a straining anxiety that did not leave me for days afterwards. It had started at the Villa of the Mysteries. To reach the villa, you exit the urban circle of Pompeii northwest and walk the monumental Way of the Tombs. Then you cross the marine dunes along the diagonal pattern of the reliefs. Only a few steps from the archeological site, the villa was left neglected, the windows open. The unparalleled fresco cycle unguarded. A couple of sagging straw curtains barely blocked direct sunbeams. The paint had faded in a spot hit by the light oozing through a gash in the weft. Visitors walked around indifferently. Anyone could destroy that priceless surface without effort nor motive. That anxiety had cast me into a transient time of uncertainty, permanent fluctuation, weather and atmosphere, currents and nimbuses.

[7]The Italian version of Tischbein's letter reads: "Il padrone della nave ha fatto un lauto bottino: ha trovato danaro e merci in quantità, seterie, caffè, e inoltre dei ricchi gioielli, appartenenti a una giovane mora. Era interessante vedere la folla che a migliaia si spingeva sui canotti per vedere da vicino i prigionieri e specialmente la mora. Vi erano anche alcuni che avrebbero desiderato comprarla…".

[8]Translation by the author.

That psycho-cosmic instant was for me the sensing of possible incommensurable losses in the plural journey of many lives, transient states, temporality, inscrutably mutable circumstances that subtend the space of landscape as opposed to the eternity of architecture.

6.2.11 Essential Continuity of Garden and Landscape

There is an essential continuity in the spatial assemblage that we have described between the constructs of the Pompeiian gardens and the maritime landscapes of the Parthenopean bays. A continuity that stands in the fluxes of matter and people that cross the waters, the energy traversing the atmospheric events, the currents that move the liquid masses of the sea, the winds that propel the sails, the succession of the Neapolitan flora and its inhabiting fauna on the shores scattered with relics and ruins, in the urban centers and in the sprawled coasts, but, also, it stands in the mythology that has been disseminated over the lands across centuries. From the city-land fringe of the western hillsides of Pompeii and the urban space of the Bay of Naples, to the extra-urban seashore of Baia Domitia in the Gulf of Gaeta, an uncontained garden is determined by a functioning metabolism and an imbued mythology.

6.3 Bartering Symbiosis for Service

At a point in time, the construct of the garden suffers a disruption. The idea of the garden as part of a broader atmospheric assemblage that interfaces man and cosmos is lost. A more utilitarian vision rises to flatten the natural components and systems harnessed by man within the built space, into services. A positivistic attitude, in fact, still today calls ecological services any natural resources useful to man. This shortsighted interpretation of the surrounding world becomes hegemonic when the period of the great explorations of the New Worlds yields to systematic research, the privileged time of the transition from Enlightenment to Romanticism yields to overt Romanticism, the formation of the image of the *patrias criollas* yields to the constitution of the *estados nacionales* in the Spanish colonies, and the cities start to equip themselves for their metropolitan metamorphosis implementing large and dull public gardens at the service of a broader number of urbanites.

6.3.1 Rousseau, Goethe, Humboldt, and the Picturesque Olmsted

In 'Rousseau, Goethe, Humboldt: Their Influence on Later Advocates of the Nature Garden', Jost Hermand identifies a line of involution that in the first half of the nineteenth century converts the Enlightened-Sentimentalist (of the Enlightenment's Age of Sentimentality) idea of the 'nature garden' into the 'vulgarized' model of the 'English garden', epitomized by Peter Joseph Lenné's design for the Volksgarten in Magdeburg of 1824, considered the first example of a people's garden in Germany (Hermand 1997).

Hermand notes that Julie's garden at Clarens in Rousseau's epistolary novel of 1761, *Julie, or the New Heloise*, is the place of Enlightened liberation from the constraints of 'absolutist' ('feudal') society and religion ('clerical chains') in harmony with, or approximating the wild growth of, a pristine state of nature, an Elysium or Arcadia, "reminiscent of the lifestyle of the Golden Age" (Hermand 1997, pp. 35–7) The Volksgarten of Magdeburg, instead, trades the ideal of individual liberation for the planning of public parks that strain the city budget as little as possible by implementing cheap "large lawns, randomly interspersed with groups of bushes and trees" (Hermand 1997, p. 45) of low maintenance.

6.3.2 Gardens and Revolution

According to Hermand, the historical turning point that "dashes all hopes of equality and fraternity" (Hermand 1997, p. 38), converting the Enlightenment into proto-modern capitalism for the maximum benefit of the upper middle class, is the victory of the Gironde of 1794, in the upheavals of the French Revolution. Known as the Thermidorian Reaction, a *coup d'état* starts with the order to execute Maximilien Robespierre issued by the Girondins' majority of the National Convention, aimed to smash the hegemony of the Montagne in the Committee of Public Safety.[9]

The earlier involution in the conception of the garden, converting a culturally sophisticated tension towards the emancipatory ideals of liberation into a practical preoccupation for the provision of some basic natural services to the urban population, is, then, the consequence of that reactionary twist in the course of the revolutionary events in France.

---

[9] In a generalized panic, generated in truth by the clash between two institutional bodies and the reciprocal suspicion among the individuals that formed them, rather than by the conflict between two political factions, the Gironde prevails. The scheme is brought forth by means of indiscriminate massacres that replace the Reign of Terror with the equally bloody regime of the White Terror.

6.3.3 Gardens and Metropolitan Aspirations

From a different point of view, in his late testament on the scenic values of the landscape, *A World with a View* (Yale University Press 1978), Tunnard identifies the implementation of people's parks, along with street widening, paving, and lighting (a broader 'municipal *vade mecum*' includes more generically a number of new avenues, malls, bridges, residential squares, terraces) as the characteristic piece of the 'metropolitan improvement' undertakings, dating from the mid-eighteenth century to become "a mark of the 19th century cities everywhere" (Tunnard 1978, pp. 127–8). What Hermand calls the vulgarized English garden, as opposed to the utopian nature garden, is in fact the mirror of a diffusive desire for 'metropolitanization' that corresponds to the structuration of industrial society.

No metropolis exists without a full erasure of nature. The metropolis is the replacement of nature with an artificial environment, which in turn implies the necessity of a man-made recreation of nature within the metropolis itself. In his retroactive manifesto for Manhattan, Koolhaas claims that the de-Witt-Morris-Rutherford Grid of 1807 announces the 'obliteration of nature' (Koolhaas 1978, p. 20) in keeping with the worldview ingrained in the fabricated motherland of the first Dutch colonists (Koolhaas 1978, p. 17).

6.3.4 Gardens and Taxidermy

In 1853, with a "colossal leap of faith" the Manhattaners foresee the not imminent but incumbent day when "the picturesquely-varied, rocky formations of the island will have been converted into formations of rows and rows of monotonous straight streets, and pile of erect buildings",[10] and an undeferrable need surges for a "taxidermic preservation of nature", where preservation stands for "manipulation [...] performed on nature saved by its designers" (Koolhaas 1978, p. 21). That is Central Park, which reformulates a "synthetic Arcadian Carpet" from "a catalogue of natural elements [...] taken from its original context" (Koolhaas 1978, pp. 22, 23). Frederick Olmsted and Calvert Vaux' plan for Central Park, otherwise a 'picturesque' landscape composition, exceeds the model of the vulgarized nineteenth-century people's park by virtue of the colossal magnitude of the enterprise and by virtue of its position in place and time, at the heart of the epochal Manhattanist edification. To Manhattan's built up mosaic, the park constitutes a fabricated otherness.

---

[10]Here, Koolhaas is quoting John Reps' *The Making of Urban America* (Reps 1965, pp. 331–9). It is a multiple-indirect quotation, since Reps' passage is, in turn, quoting Frederick Olmsted and Calvert Vaux' project statement 'Description of a Plan for the Improvement of the Central Park', registered in the archives of the Commissioners of Central Park of 1858, as published in Frederick Olmsted Jr. and Theodora Kimball's book about Olmsted (Olmsted Jr. and Kimball 1928).

6.3.5 Gardens, Effects, and Deep Forms

Olmsted and Vaux' project statement reported by Reps proceeds as follows:

> "Then the priceless value of the present picturesque outlines of the ground will be distinctly perceived, and its adaptability for its purpose more fully recognized. It therefore seems desirable to interfere with its undulating outlines, and picturesque, rocky scenery as little as possible, and, on the other hand, […] to increase and judiciously develop these particularly individual and characteristic sources of landscape effects." (Reps 1965, p. 336)

The perception of the site as a 'scenery' and the conception of the project as an enhancement of the sources of 'landscape effects' are explicitly stated. The premises set out by Olmsted and Vaux actually coincide with landscape urbanism's claims in favor of the use of the *simulacrum* as a design tool, contained in its 'manual for the machinic landscape' (Hight 2003, p. 27).

However, those premises differ from the sense that the term 'scenic' acquires in Tunnard's *A World with a View*. Tunnard's 'scenic' refers to the overall ensemble of deep forms as perceived in its truth by the beholder, while the 'picturesque *simulacrum*' refers to the perception of a scenery orchestrated in its surface characteristics to induce a desired effect.

In fact, Tunnard uses Paul Schultze-Naumburg's *Die Gestaltung der Landschaft durch den Menschen* of 1916 to lay the bases of an idea of scenery inherent to the deeper structure of landscape. In his study, the classification of a series of landscape cases "by type and detail" cross-relates man-made and natural components (Tunnard 1978, p. 119). Where the industrial territories "possess confused, vague traits", Schultze-Naumburg proposes the mass of the town of Tubingen, "built on a slope and seen from an opposite slope", as the virtuous example of a settlement situation that combines clear recognizability and harmonic merging within a context. A man-made entity part of the landscape when seen from a distance, Tubingen's built ensemble frames views of that landscape when the observer is led into its core (Tunnard 1978, p. 121).[11]

Among pioneering researches on the 'scenic values' of landscape, Tunnard mentions a romantic-picturesque field and a semi-scientific field. On the one hand, in *Italian Townscape* of 1963, Hubert de Cronin Hastings drifts towards an appreciation of the 'picturesque effect', rather than deeper compositional forms, by deploying 'romantic' categories of landscape analysis such as 'invitation', 'convexity', 'precipitation', 'personality', 'action', 'inaction', 'remoteness' (Tunnard 1978, pp. 122–3). On the other hand, Ian McHarg pursues an objective classification of the 'scenic values' of the landscape.[12] His principle of the 'least social

---

[11]Tubingen is, thus, as alternative to the industrial territory as to the man/nature divide advocated by the American system of National Parks for the preservation of nature.

[12]Founder of the Landscape Design Program at the University of Pennsylvania, McHarg's work would be used as a founding platform for the landscape urbanist movement.

cost' concerned with the identification and protection of scenic potential and cultural heritage would experience a particular success, being adopted in the highway building programs worldwide. According to Tunnard, however, McHarg's model eludes the issue of 'quality' (Tunnard 1978, p. 123).

It is, instead, the *Vedutisti* that, according to Tunnard, still offer the most efficient model for the investigation of the 'total landscape', where the city represents just 'a denser part' of a *continuum*, composed according to 'monumentality', 'relation to topography', 'architectural quality', 'visual alignments', 'views' (Tunnard 1978, pp. 123, 126–7). Tunnard's *rappel à l'ordre* for the reestablishment of "the lost principles of urban design" (Tunnard 1978, p. 123) extended to the total composition of the architecture-topography continuum and made of 'monumentality', 'view', 'architecture quality' (Tunnard 1978, p. 127), recall Aldo Rossi's disciplinary refoundation. Even remaining prevalently culturalist,[13] Tunnard's 'scenic' advocates for the combination of a "scientific approach to nature" (Tunnard 1978, p. 182) with the "full development of the cultural patrimony" (Tunnard 1978, p. 185) pursued through panoramic and detailed observation. Tunnard's method, in sum, prescribes a twofold investigation of the deep structure of the *continuum*, building knowledge of both physical landscape development and landscape imageability precipitated across its long-term cultural recognition. The scenic values are, thus, inherent to the deep forms, that is, structures, of the landscape, not its effects.

We could eventually induce that Central Park and the High Line are closer to Cony Island's 'technologies of the fantastic', (Koolhaas 1978, p. 29–79) or to a dissimulated standardized, industrialized subversion rather than an emancipatory agenda for the liberation from feudal hierarchization of the world claimed to the revolutionary nature garden by Hermand.

6.4 Game Hunting in the Garden

The gradual configuration of Berlin's Tiergarten can be broken down in three phases. First, its original establishment at the beginning of the sixteenth century corresponds to the delimitation of the exclusive hunting grounds of the Great Elector of Brandeburg, where a surrounding fence prevented the escaping of wild animals from the inside and undesired incursions of poachers from the outside.

[13]Culturalist is, in fact, Schultze-Naumburg's idea of man/nature symbiosis on which Tunnard elaborates: "anything manmade should be a harmonious part of the landscape", aiming at the restoration of the original "symbiosis [...] undone in the second half of the nineteenth century" by economic circumstances winning out the "understanding of nature's beauty" (Tunnard 1978, pp. 119–20).

6.4.1 Metropolitan Services Under the Tents

Second, the 'urbanization' of the *garten*, that is its intertwining in the bundle of expanding urban dynamics, can be traced back to the time of the extension of the Unter den Linden Boulevard to connect the Stadtschloss to the new Schloss Charlottenburg. At the order of Friedrich Wilhelm I, the central Grosse Stern roundabout and a pattern of radial boulevards are implemented at the turn into the eighteenth century. The conversion of the hunting park to *lustgarten*, or garden of pleasures, open to the public, is completed in the 1740s under Friedrich II along the lines of Georg von Knobelsdorff's plan, which equips it with the functional and ornamental features of the French park. There appear pathways, flowerbeds in geometrical layout, greenery salons, mazes, water ponds, and fountains. The access is unrestricted. The park is thus regarded as a service to the urbanites, who by 1745 sense the opportunity of refreshments served under seasonal tents installed in a circus. It is a circular esplanade on the northeastern limit of the area by the Spree River bank, from which eight radial avenues depart to disappear in the forested park. The concession for the refreshment tents is granted to Huguenot refugees, whose heirs by 1786 can edify a permanent construction starting the consolidation of a small urban ensemble.

6.4.2 Vulgarized Beautifications

In the third phase of the park's evolution, in fact, Lenné's intervention formalizes the presence of the refreshment facilities by beautifying their urban front with a lofty double-tree-lined street, the In den Zelten. The elegant establishments would stand until the Second World War bombings.

Lenné's intervention, largely realized in the 1830s upon commission from Friedrich Wilhelm III, is generally intended for the beautification of the park, by incrementing the number of major and secondary pathways over Knobelsdorff's general matrix. Forest swathes are characteristically reduced to clear open lawns traversed by ornamental water features and spotted with shady clusters of trees. The transformation of the Tiergarten in accordance with the model of the nineteenth-century English park is completed with the creation of the Zoological garden in 1844, both popularized curiosity grounds and educational facility for the growing population of a city in the process of metropolitanization.[14]

---

[14]The evolution of Berlin's *Tiergarten* does not differ much from that of Hyde Park in London, created as Henry VIII's hunting grounds in 1536. The early eighteenth-century geometric gardens of Kensington Palace were split in two parts with the introduction of the Serpentine River and Long Water by Henry Wise and Charles Bridgeman at the request of Queen Caroline in 1728. West lies the formal design of the Kensington Gardens, an Italian garden exhibiting a Dutch garden core, east is the metropolitan park in the service of the urbanites. Hyde Park, admired by the future Napoleon III in the years of his second exile, becomes the model for the implementation

6.4.3 Emancipation or Ecological Services

Lenné's transformation of Knobelsdorff's Tiergarten scheme in Berlin, is representative of the vulgarizing attitude lamented by Hermand for the Volksgarten of Magdeburg. The conversion of Rousseau's aspirations to a natural idyll into the plane provision of ecological services for the urbanites is, in fact, the apotheosis of the anthropocentric vision of landscape that characterizes metropolitanism.[15] But in more general terms, we can regard the historical transition from the *tiergärten* and hunting gardens to the metropolitan parks in the European capitals as the perpetuation of the feudal structures of the Middle Ages into modernity.

The metropolitan model, once more, propounds a predatory attitude towards nature that results incompatible with an advanced phase of civilization, aware of the overall metabolic balance in the ecosphere, and, even more, in the face of the unsustainably growing human population on the planet. The idea of big game hunting can be replaced today by more civilized forms of leisure, as its globalized modern correspondent, the exploitation of natural services, can be replaced by an alternative ethos.

6.5 Hunts of a Subtler Sort

6.5.1 Souvenirs Entomologiques

Jünger's *Subtile Jagden* (subtle hunts) of 1980 are probably inspired by Jean-Henri Fabre's *Souvenirs Entomologiques*,[16] a collection of essays and memoirs of the French entomologist, known for disregarding the Darwinian theory of evolution of the species and focusing his natural observations on the behavior of animals alive in the wild instead of the anatomical dissection of their dead bodies in the laboratory. The emancipation from a positivistic analysis of nature takes advantage from two principles that characterize Fabre's work: the prevalent consideration dedicated to the relational dynamics and the inclusion of the autobiographic data of the observer as part of the system.

---

of the Bois de Boulogne and Bois de Vicennes in the frame of the urban renovation of Paris during the Second Empire.

[15]Landscape is identified with a provision of ecological services just like woman has been traditionally regarded by man as a provision of natural services. See, to this regard, the comparison between the male bachelor of Koolhaas' Manhattanism with the female nomad of Ito's Tokyoism in Chap. 2 of this book. Also see John Berger's *Ways of Seeing* about female immanence (Berger 1972). Even today, nature is widely termed ecological services. Deleuze Guattari's 'becoming-woman' and 'becoming-animal' processes advocate for a collective emancipation from this anthropocentric, and androcentric, movement of history across geography.

[16]See Fabre's series *Souvenirs Entomologiques. Étude sur l'instinct et les moeurs des insects*, 1879–1909.

Fig. 6.2 Glass negative portraying Jean-Henri Fabre at his desk in the Sérignan-du-Comtat house, around 1910 (*image* Bain News Service, Bain News Service, PD legal notice at end of chapter)

In fact, the *Subtile Jagden* closes with an account of Jünger's visit to Fabre's museum-house of Sérignan-du-Comtat in Provence. Jünger was certainly equally interested in the observation of the entomological form as the entomological behavior, but his interest in natural dynamics, the system of relations among elements in the wild, is poetically narrated as the discovery of temporary fragments of a universal harmony.

The assemblage of the poor instruments, box cutters, tweezers, and spatulas, arrayed on Fabre's desk, along with the names, dates, books, titles, and various objects he had once handled in the museum-house, is associated with mortuary equipment consumed by the unrestrainable decay of time and light (Fig. 6.2).

6.5.2 *Gardens Grown Wild*

In sharp contrast, the effects produced by the passing of time in the garden are quite the opposite. The garden has grown wild, occupied by the Mediterranean *maquis*, but that process has brought into it a new buzzing life of bees, lizards, birds that dart crying through the bushes, herbs, and mosses. A life 'forgetful of all scientific names' where 'things speak with their force' and even the dead master resurrects from his museum-tomb with the surrounding nature, in the quiet of the full sun (Jünger 1997, pp. 271–2).

Jünger's description of Fabre's garden is, historically, in line with the *Heimatschutz* league's call for the transformation of small bourgeois gardens into nature gardens by letting grow wild at least a corner of them (Hermand 1997, p. 50).[17] But it is also reminiscent of Saint Francis of Assisi's precepts to his Orders, recommending that a 'part of the friary garden always be left untouched'.[18] By mirroring the beauty of the universe in wild flowers and herbs, a garden's corner grown wild affords a direct channel between material reality and supreme harmony.

6.5.3 Subtle Hunts

In *Subtile Jagden*, Jünger describes his passion for the entomological investigation that he calls 'subtle hunts' as opposed to 'big game hunting'. The book is the autobiographical account of his colorful explorations across the globe driven by entomological and botanical curiosity. The subtle hunts of the *scientiae amabiles*, entomology and botany, are for Jünger a gymnasium to train the modes of the spirit for the comprehensive observation of nature. The dynamic microcosmos of the insects is the metaphor of the macro-cosmos.

As Alessandra Iadicicco notes in her short introduction, Jünger needs not to refer to the sidereal distances of the asters to enjoy the plurality of the worlds, since a perspective over the minuscule world of insects, prodigious creatures endowed with the same organs as we do, and even winged, can shed light over their mysterious relationship with the *infinitum*.

6.6 Chapter's Conclusions

6.6.1 Darwinism and Technique

Maria Bellucci observes that, in *Subtile Jagden*, Jünger critiques Darwinism's leaning towards utilitarianism as an analogue to the implacable expansion of technique and statistics that uniforms and, by that, desertifies the modern world

[17]See also Wilhelm Bölsche's *Die and Become* of 1913.

[18]"For this reason, [Saint] Francis asked that part of the friary garden always be left untouched, so that wild flowers and herbs could grow there, and those who saw them could raise their minds to God, the Creator of such beauty" (Francis 2015, p. 11).

(Bellucci 2015). Jünger, in fact, stresses that the technical equipment that facilitated Alexander von Humboldt's geographic observations of the Orinoco River never impeded his comprehensive gaze, nor prevented his perception of the earth's pulse.[19] Contrariwise, Darwin's gaze is oriented by the very concept of utility that contributes to his profiling of creatures and their behavioral traits while impoverishing them.

Jünger laments that the same simplifying duality of 'love' and 'war', or 'coupling' and 'struggle for life', that structures the Darwinian theory resonates in the simplification of the industrial landscape and its few, reduced types shaped for easy use (Jünger 1997, p. 109). In the industrial landscape, whose production characterizes our epoch of technique, the simplification of building types appears along with the rapid reduction of biodiversity and 'a transformation of form hides behind biological transformations' (Jünger 1997, p. 110).

According to Jünger, mathematical calculus and the big numbers of statistics insinuate themselves into the observation of nature through the efficient methods of industrialized research, drying up the contemplation of the beholder and making even the entomological determination of species 'a matter of sophistication' (Jünger 1997, pp. 168–9). 'Economic desertification and its methods' parallel the rise of industrial research that in teamwork studies sophisticated poisons to be sprayed onto enemy trenches as well as over the extended regions of industrialized agriculture, killing plagues along with wild fauna, and, by that, 'terminating the time of subtle hunts' (Jünger 1997, p. 270).

6.6.2 *Prey Mirrors Hunter*

'The ancients knew that prey is the mirror that ends up shaping the hunter'. And the hunter's soul is given form by the subtle modes of his unflagging hunt, becoming eventually as subtle as the immaterial waft of the wings of the ephemerides and the other minuscule preys he has been chasing after. (Jünger 1997, p. 270). Technique is therefore terminating the hunter along with the prey.

Jünger's message is that the hunter has to undergo a metamorphosis not to be terminated by his own action over the planet. The idea of big game hunting can make way to much subtler hunts, that is, the exploitation of natural resources can be replaced by an action of observation and understanding of the world and its relational, that is, symbiotic, assemblages, and dynamics.

---

[19]In his essay accompanying the edition of von Humboldt's *Geography of Plants*, Stephen Jackson gives a detailed description to the 'Instruments Utilized in Developing the *Tableau physique*' in comparison to those used by Horace-Bénédict de Saussure for scientific measurement during his explorations of the Alps a couple of decades before (Jackson 2009, pp. 221–6).

References

Bellasi P (1987) Il giardino del pelio. Segni, oggetti e simboli della vita quotidiana. Costa & Nolan, Genova

Bellucci M (2015) Book review. Naturalmente, Fatti e Trame delle Scienze

Berger J (1972) Ways of seeing. BBC and Penguin Books, London

Ciarallo A (2002) Il giardino pompeiano. Le piante, l'orto, i segreti della cucina. Electa Napoli, Napoli

Ciardiello R (2006) Insula occidentalis 42. In: Aoyagi M, Pappalardo U (eds) Insula Occidentalis (Regiones IV–VII). Valtrend Editore, Napoli

Deleuze G, Guattari F (1987) A thousand plateaus: capitalism and schizophrenia (trans: Massumi B). The University of Minnesota Press, Minneapolis [1980]

Edwards C, Woolf G (eds) (2003) Rome the cosmopolis. Cambridge University Press, Cambridge

Foucault M (1984) Of other spaces: utopias and heterotopias (trans: Miskowiec J). Architecture/Mouvement/Continuité 5:46–49 [1967]

Francis P (2015) Encyclical letter laudato si' of the holy father francis on care for our common home. Vatican Press, Vatican City

Goethe JW (1993) Il viaggio in Italia. In: Santoli V (ed) Goethe: opere (trans: Zaniboni E) Sansoni, Firenze. [1813–7]

Hermand J (1997) Rousseau, goethe, humboldt: their influence on later advocates of the nature garden. In: Wolschke-Bulmahn J (ed) Nature and ideology. Natural garden design in the Twentieth Century. Dumbarton Oaks, pp 35–57

Hight C (2003) Portraying the urban landscape: landscape in architectural criticism and theory, 1960-present. In: Mostafavi M, Najle C (eds) Landscape urbanism: a manual for the machinic landscape. Architectural Association, London

Jünger E (1997) Cacce sottili (trans: Iadicicco A). Guanda, Parma [1980]

Jackson ST (2009) Instruments utilized in developing the tableau physique. In: von Humboldt A, Bompland A (eds) Essay on the geography of plants. The University of Chicago Press, Chicago, London, pp 221–226 [1807]

Olmsted Jr F, Kimball T (1928) Frederick law olmsted. New York

Rem K (1978) Delirious new york. Monacelli Press, New York

Reps John (1965) The Making of Urban America. Princeton University Press, New York

Tischbein W (1993) Lettera di Tischbein. In: Santoli V (ed) Goethe: opere, Il viaggio in Italia (trans: Zaniboni E). Sansoni Editore, Firenze [1813–7]

Tournier M (1997) Gemini (trans: Carter A). The Johns Hopkins University Press, Baltimore [1975]

Tunnard C (1978) A world with a view. Yale University Press, An Inquiry into the Nature of Scenic Values. New Haven

Legal Notice:

Chapter 7
Reconciling Cultural and Cognitive

Abstract This chapter dissects a contemporary idea of landscape into its multiple components. The text analyzes a series of landscape constructs in a conceptual sequence moving from elemental to elaborate: *ur-Landschaft*; habitat; environment; territory; cultural landscape; cognitive landscape. The chapter discusses various positions that have been recently confronting over the notion of landscape, whose alternative landscape paradigms gather around two opposite fronts: the 'culturalist-geophilosophical' and the 'scientific-cognitivist'. The text elaborates on a number of the diverse layers and components referable to a contemporary landscape notion. Components and layers derive from literature and practice, originating in the present and in the recent or remote past. The text, then, examines with specific focus the contended character of continuity or discontinuity of the notion of landscape and its evolution both in time and space, on which the very essence of the notion is dependent. The text compares positions describing the landscape as an aboriginal construct generated by the man/nature interaction and others regarding landscape as a sophisticated construct only generated when man's manipulation of space drops any utilitarian purpose aiming at a greater perfection. The text then construes a landscape paradigm capable to accommodate both culturalist and cognitivist layers and components from previously analyzed models. Reformed and recombined the layers and components conform a new 'symbiotic landscape paradigm' that is described and graphically represented in a diagram.

Keywords Landscape paradigm · *Ur-landschaft* · Habitat · Environment
Territory · Cultural landscape · Cognitive landscape · Metabolist
Geophilosophical · Symbiotic

7.1 Introduction

The concept of 'landscape' in several different conjugations has recently gained concurrent centrality in the discourse on architecture as well as aesthetics, showing surprising osmosis while consolidating the role of autonomous disciplinary branches in both fields: landscape and environmental design in the former and landscape aesthetics in the latter (Fig. 7.1).

R. Pasini, *Landscape Paradigms and Post-urban Spaces*, The Urban Book Series,
https://doi.org/10.1007/978-3-319-77887-7_7

Fig. 7.1 Imperial German Army Lieutenant Ernst Jünger portrayed with the Pour le Mérite, the Order of Bravery instituted by Friedrich the Great, awarded to him in 1918 while convalescing from the last of multiple combat wounds received in the World War I battles of Hauts-de-France. Jünger's intense countenance is framed by the large fur collar of the military coat. The image condenses the idea of a relationship between man's interiority and the landscape represented by the hair, reminiscent of the shaking boughs of the woods and the animal world (*photo* courtesy of Deutsches Literatur Archiv Marbach)

7.2 Breaking Down the Idea of Landscape

7.2.1 Fundamental Components

In his entry *Estetica ambientale* (2010) in the *Enciclopedia Treccani*, Paolo D'Angelo focuses on the recent interest in 'environmental aesthetics' and 'natural beauty', identifying a confrontational positioning of the research milieu around a 'scientific-cognitivist' model versus a 'geophilosophical' and 'atmospheric' model. D'Angelo identifies the object of interest of the former model with the mechanics of the environment, a physical-biological apparatus, and attributes to the latter exclusive competence over the landscape proper, as a relational-cultural construct.[1] The shortsighted juxtaposition of 'philosophy', which the discourse on the landscape supposedly pertains to, versus 'ecology and architecture', regarded as improper interpreters of the theme,[2] ends up ostracizing valuable contributions on one front or the other. In fact, the orchestration of paradigms proceeding from different disciplinary backgrounds shows great potential for tackling the novel challenges posited by the extended notion of landscape.

The use of philosophical paradigms to inspire, support, or experiment new design approaches has become quite common among architects producing both earnest and specious operations. But the architects' frequentation of the field of landscape has also proved useful to philosophers and aestheticians.

In his contribution to *Natural Histories*, the catalog of the exhibition on Herzog and de Meuron's models and material archives curated by Philip Ursprung, Gernot Böhme develops the concept of 'atmosphere as the subject matter of architecture' (Böhme 2006, pp. 398–406). Atmospheric architecture is, in fact, a derivation of his more general paradigm of 'atmospheric landscape'. Böhme situates architecture in the modulation of perceptive subject/surroundings interaction. A subject explores space through 'bifocal vision', 'movement and focusing function', and completes it by sensing a 'mood' for taking part in the scene. A relational *locus* of 'contiguity and surroundings' and a measurable *spatium* of 'distance and scale' are thus subjected to multisensorial 'concentrations', 'orientations', 'articulations', and 'deformations' due to the "incorporation of the subject's physical presence" and the experienced 'atmosphere' (Böhme 2006, p. 403). As noted by D'Angelo, Böhme's assimilation of architecture with the design of a larger stage, in the sense of a

[1]"The 'environment' is a physical-biological concept, while the 'landscape' is a relational concept, which has to do with the way in which we represent a territory and we feel in it […]. The geophilosophical and the atmospheric models recover substantial parts of the history of the landscape" (D'Angelo 2010a) [Translation by the author].

[2]"The discourse on the landscape seems to have migrated towards other disciplines, such as Architecture and Ecology […] The landscape is, however, a specific object of philosophical reflection […] in consideration of its aesthetic dimension" (D'Angelo 2010b) [Translation by the author].

'phenomenological landscape', draws on both the romantic science of nature (Goethe, von Humboldt, and Carus) and Husserl's 'intersubjective phenomenology' (D'Angelo 2010a). However, the derivation of Böhme's atmospheric paradigm is clearly Nietschian with reference to the "vaporous region of the unhistorical" that is "like an atmosphere where life alone can germinate" (Nietsche 1997, pp. 63–4). The notion of the vaporous unhistorical passes over to Böhme via Deleuze Guattari, who reformulates it into the idea of "geography as opposed to history" (Deluze and Guattari 1987, p. 296) to lay the foundations of geophilosophy.

7.2.2 Atmospheric Perfumes

Ursprung's introduction to the chapter *Beauty and Atmosphere* in *Natural Histories* describes Herzog and de Meuron's projects as "bursting in space like iridescent soap bubbles" (Ursprung 2006, p. 364). Ursprung's metaphor then unfolds in a parallel between 'architecture' and 'scents'. Both architecture and scents can be regarded as libraries of memory that store the materials collected via the subject's physically involving experience. In these records of subjective materials lies the faculty of the 'aura of a place' to conjure up a unique identity.

Ursprung's description certainly sounds reminiscent of Saint-Augustinian mnemonic repositories of available emotions and perceptions, tending to favor the individual over the intersubjective dimension, the spiritual over the social and structural. Saint Augustine says

> And I come to the fields and spacious palaces of my memory, where are the treasures of innumerable images, brought into it from things of all sorts perceived by the senses. […] There are all things preserved distinctly and under general heads, each having entered by its own avenue: as light, and all colours and forms of bodies by the eyes; by the ears all sorts of sounds; all smells by the avenue of the nostrils; all tastes by the mouth; and by the sensation of the whole body, what is hard or soft; hot or cold; or rugged; heavy or light; either outwardly or inwardly to the body. All these doth that great harbour of the memory receive in her numberless secret and inexpressible windings, to be forthcoming, and brought out at need. […] For even while I dwell in darkness and silence, in my memory I can produce colours. […] Yea, I discern the breath of lilies from violets, though smelling nothing; and I prefer honey to sweet wine, smooth before rugged, at the time neither tasting nor handling, but remembering only. (Augustine 1909, pp. 173–4)

Analogies between Herzog and de Meuron's designed repositories of phenomena and Aldo Rossi's idea of the 'individuality of urban artifacts' relying on the subject's experience and memories of a place are also as obvious as largely disregarded by critics. In El Croquis n. 152–153, the editor Jean-François Chevrier points out the atmospheric ephemerality of Rossi's *Teatro del Mondo*, a paradoxical piece in the work of an architect who has always advocated for monumental permanence. The ephemeral character of the *Teatro* can be explained in the terms of a phenomenological, atmospheric receptivity in capturing "water, air, the elements […] and light" of the Venice lagoon, transforming architecture in a "*locus* for life and experience" (Chevrier and Herzog 2010, p. 39).

7.2.3 Using the Philosophy of the Landscape to Do Architecture

Chevrier's interpretation of Herzog and de Meuron's recent work pivots on this 'atmospheric paradigm' referable to Böhme and to the philosophical reflection on the landscape. The result is that of transferring any problem of architectural composition onto the larger field of landscape. Herzog and de Meuron's claimed resourcing to the "landscape [to replace even] context, ecology, nature" (Chevrier and Herzog 2010, p. 40) in the contemporary city is evident in their latest architectural practice.

Abstinence from the architectural object's formalization is intentionally pursued[3] in favor of an experiential assemblage of social activities and landscape materials that conjure up a formless space. The adoption of the 'landscape' paradigm is explicitly formulated in recurrent statements for the demise of 'style',[4] in favor of a perceptual multiversity of architecture in accordance with the 'complexity of nature' (Chevrier and Herzog 2010, p. 22). Chevrier's interpretative formula for Herzog and de Meuron's projects consists in what we could call a 'hyper-urbanist' structure with 'sub-urbanist' opening, where the 'place' is produced by the combination of the 'program's activities', while that program's visibility is organized over the structure of the manipulated landscape (Chevrier 2010, p. 8).

7.2.4 Ur-Materials

The 'landscape' model applies to the composition of the Plaza de España project in Tenerife, where a monumental roundabout plaza on the neoclassical waterfront is converted into an atmospheric ensemble recomposing the '*ur*-materials' of the geological generation of the volcanic island. Jacques Herzog openly words it: "the Ocean is three thousand meters deep […] the summit of the Teide stands four thousand meters above its surface. The volcanic emergence of the island is a sculptural gesture of inexpressible violence" (Chevrier and Herzog 2010, p. 41).

---

[3]"The working method has become a form […]. It is a method of working that has slowly become architecture", Jacques Herzog's words (Chevrier and Herzog 2010, p. 33).

[4]"We wanted to avoid style: the idea of perception is more open", Jacques Herzog's words (Chevrier and Herzog 2010, p. 23).

Countering the rhetoric of a Francoist stele, a water basin is conceived as both a “flooded town square” and a “small shallow lake”. The brink of the depression draws a white sandy shore in the dark paving with the texture of lava flows, while “glass bubbles resembling water drops” are suspended for lighting, and elusive small buildings are covered with ferns like rocks with musk and lichens. The meager foliage of slow growing new plants establishes a divarication with the abundance of the preexisting trees inscribing that “microcosmos into a temporal dimension” resonating in landscape succession (Chevrier 2010, p. 14).

The manipulation of geographic elements in the Plaza de España reformation acquires an exemplary stature for its location at the foot of the majestic Pico de Teide. Halfway in the Atlantic between Europe, Africa and Amercia, the 4000-m-high Teide is the first summit climbed and observed by Alexander von Humboldt and Aimée Bompland at the beginning of their legendary equinoctial expedition.[5] The ascent works for them as a rehearsal for the legendary Chimborazo ascent and a field test for their new scientific instruments.

7.2.5 *‘North-Sea/Mercantile-Mobility/Water’*

The Elbphilarmonie in Hamburg constitutes a grander example of a composition at the territorial scale, where the landscape is vertiginously transcribed “into the vertical dimension of an enormous optical machine” (Chevrier 2010, p. 19). The ‘multifunctional complexity’ of a daringly articulated program is allocated according to the “(multisensorial) capacity of the human body immersed in the world”, reaching up as far as to the production of metaphors’ (Chevrier 2010, p. 21).

The architectural ensemble reconstructs a possible terrain for the exploded Hanseatic landscape. The artificial massif of the existing docking warehouse Kaispeicher A, at the very tip of the peninsula, is used to lift onto a podium the immense crystal of the programmatic assemblage that encapsulates the auditorium. As a territorial mirror, the crystal “captures and combines reflections from the sky, the water and the city”.[6] The undulating roofline rising up to +110 m, reflects the waves of the North Sea as well as the curves of a nomad tent.

An articulated program of civil sociality and intersubjective play, is distributed over the vast terraced platform contained in the profound fracture in between the

[5]The island of Tenerife is reached in June 19, 1799 by the Spanish corvette Pizarro carrying the expedition. The ascent is carried out two days later, testing for the first time the sophisticated instruments commissioned to the most distinguished toolmakers of the time in London, Paris, and Geneva.

[6]See: Project 230 Elbphilarmonie Hamburg, Herzog and de Meuron’s website.

opaque podium and the translucent crystal. The macro-composition is thus legible as a landscape construct vertically stacking assemblages of '*ur-Landschaft*/geology/rock', 'North Sea/mercantile mobility/water', with the web of 'sociality/sedentary city/air in between and throughout'.

7.2.6 Elective Belonging: Using the Architecture of the Landscape to Do Philosophy

The 'geophilosophical' landscape paradigm of 'elective belonging' elaborated by aesthetician Luisa Bonesio is equally founded on Deleuze Guattari's geophilosophy as well as on the decade-long work of the *scuola territorialista* of architect and planner Alberto Magnaghi, which Bonesio extensively cites and credits (Bonesio 2002, p.6). Magnaghi's vision of territorial heritage is based on a "structure of invariant formal matrices producing durable wealth"[7] as well as on their recognizability and availability for adoption by non-indigenous agents (Magnaghi 2010, p. 299).[8] The opening to elective belonging tempers the regressive Heideggerian roots of geophilosophy. However, the territory is a platform of interaction, intersection, and coexistence between *ur*-platform, the natural platform of the origins, and the great assemblage of the structural and infrastructural networks, the intangible fluxes and the social, economic, and geopolitical positioning, juxtapositions, and dynamics. In the case of the geophilosophical model of the elective belonging, in fact, the work of the *scuola territorialista* under-arches an aesthetic reflection and an idea of territory supports a cultural interpretation of the landscape.

7.2.7 Metabolism, Tempered Cognitivism, and Shape-Giving

The environment is a fundamental component of the 'expanded landscape' as are its aesthetic and territorial layers. Peter Baccini and Paul Brunner's 'metabolism of the anthroposphere' aspires to calculate the flux and exchange of matter, energy, populations, and dynamic activities through a defined field (or ecosystem) in combination with cultural and institutional interaction (Baccini and Brunner 2012).

---

[7]Translated by the author.

[8]In fact, the individual's belonging to a locale and a community does not exclusively depend on the place of the individual's birth, the *ur-Land*, generating an immutable bond, but rather on a conscious election that can occur at a later moment of the individual's life.

The concept of 'metabolism' is fundamentally centered on a quantitative engineering calculus, drifting by nature towards a form of scientific determinism.[9]

In *Grounding Metabolism*, Daniel Ibañez and Nikos Katsikis reformulate the metabolist model[10] while aiming to define some form of intentionality (albeit collective or impersonal) implementing the shape-giving process on a regional scale. They explore the material configuration of space as a "geographical imprint" of intangible "metabolic processes" (Ibañez and Nikos 2014, p. 3). But in contrast to any "metabolic determinism in which conditions on the ground are seen as a mere reflection of metabolic processes", grounded metabolism "interprets design as a geographic agent [and is] focused on the physical configuration of human occupation on the ground" (Ibañez and Nikos 2014, p. 6). The announced purpose of reconciling the world-making agency implicit in design activity with the metabolic determinism implied by its contemporary geographic scale points out a substantial theme of the contemporary debate, which is then developed in convoluted rhetorical plots. Pointing towards the same direction, Erle Ellis' radical description of the replacement of 'biomes' (natural landscapes) with 'anthromes' (anthropogenic landscapes) over the surface of the planet[11] addresses non-catastrophic, progressive prospects (Ellis 2014b, pp. 21–7) with a more effective and sober logic. If the anthropic impact on nature extends in proportion with the sociocultural and technical capacity of increasing productivity as well as environmental control, according to Ellis, a necessary novel role for design is prefigured in the 'stewardship of global sustainability'. That is intended as a non-definitive, but rather constantly adjusting, management of the metabolisms of the planet, or anthropic ecosphere, based on realistic technological know-how, with concurrent aesthetic and ethic objectives and consequence.[12]

---

[9]Discussing of Conrad Waddington's legacy, Hadas Steiner concludes that the notion of 'environment' augments that of 'habitat' by expanding the sphere of the immediate interactions of an organism to a larger ambit, where interchanges can be prevalently intangible, based on the passage of information, its processes and its networks. The 'environment', then, contains the idea of a 'cybernetic interchange' between organisms and their context, opening the way to bio/technological hybridization (Steiner 2014, p. 89).

[10]"In a condition of socio-environmental transformation […] urban metabolism [suggests] an analytical basis for gauging the continuous flows of energy, material, and population exchange within and between cities and their extensive operational landscapes" (Ibañez and Nikos 2014, p. 3).

[11]"In the Anthropocene, there is no possibility of removing human influence from ecosystems: anthropogenic transformation of the terrestrial biosphere is essentially complete and permanent" (Ellis 2014a, p. 180). See also illustrations 'a' and 'b' (Ellis 2014a, pp. 170–171).

[12]"We have never had more power to do great things, to design better landscape ecologies both for sustenance and for nature, to create beauty, and to manage a biosphere that will nurture, please, and honor our children, ourselves, and our ancestors" (Ellis 2014b, pp. 180–181).

7.2.8 *Ur-Landschaft, Territory, Landscape*

Referring to the notions that we touched upon at the beginning of the book, we can reconstruct a conventional conceptual progression leading from '*ur-Landschaft*', or pristine natural site, to 'environment', or natural system with intelligible mechanics, and 'territory'. We can assume that the latter term, territory, has been knowingly referring to the physical confluence and metabolic interaction between that natural platform and the superimposed anthropic layers since when its Renaissance equivalent '*paese*' is put in use, as Piero Camporesi points out.[13]

According to Camporesi it is the "convergenza sinergica tra operosità creative e visualizzazione della realtà",[14] detectable in the work of the mid-fifteenth-century Tuscan technical treatisers, such as Francesco di Giorgio Martini, Antonio Averlino Filarete, and Vannoccio Biringuccio, that denotes the occurrence of a common 'way of regarding nature and reading the landscape' shared by an 'entire cultural milieu'. The same perception of the surrounding landscape is reflected 'in the eye of the painter, the architect, the sculptor, as well as in the eye of the natural philosopher, the metal prospector, and the mining engineer'[15] (Camporesi 2016, p. 23).

A 'minute casuistry' of the 'landscape survey and reading' accumulates through the work of the traveling experts searching for the hidden veins of mercury, tin, copper, iron, lead, silver, sulfur. 'Every vein of minerals hidden underground leaves particular signs on the surface, soil and rocks variously colored', so that the reconnaissance of the landscape of the surface is interfaced with the fathoming of the geological profundities. The 'masters of the foundries' and the 'masters of the brush' converge into 'a common research focused on the colors, forms, and light qualities'.[16] The magnificent landscapes, backdrops to Mantegna's *Crocifissione* and Leonardo's *Vergine delle rocce*, are the collective achievements of that community of '*artieri*' (artists/craftsmen) equally, and prior, keen on the geological stratigraphy and 'mineralogical image of nature' as on its superficial manifestation. (Camporesi 2016, p. 21) It will be the Flemish and German masters of the 15th and 16th centuries, the van Eyks and Dürer, who complete the work of the '*artieri*' by "stripping nature its veil and stealing its true image" (Burckhardt 1876, p. 32).

---

[13]"Nel Cinquecento non esisteva il paesaggio, nel senso moderno del termine, ma il paese, qualcosa di simile a quello che per noi è oggi il territorio o, per i francesi, l'*environnement*" (Camporesi 2016, p. 5). "In the Cinquecento the landscape, in the modern sense of the term, did not exist, but rather the *paese*, something similar to what today is to us the territory or, for the French, the *environnement*." [translation by the author]. Camporesi notes how, in the mid-sixteenth century, fray Leandro Alberti is already familiar with the term '*territorio*', which he broadly disseminates through his *Descrittione di tutta Italia* (Camporesi 2016, p. 166).

[14]"the synergic confluence between creative industriousness and visualization of reality" [Translation by the author].

[15]Freely paraphrased by the author.

[16]Freely paraphrased by the author.

The 'implacable observation of nature' furiously practiced by Leonardo is the highest model of the alliance of 'aesthetic interest' and 'scientific curiosity' that produces the notion of the Italian landscape in the Quattrocento (Camporesi 2016, pp. 24–5). Arisen in a still semi-ancient age, only later, that notion will evolve into the Northern-European painting of the seventeenth-, eighteenth-, and nineteenth centuries, in which the landscape moves from background to foreground.

It seems consistent to conclude that there is a notion of landscape that comes directly from the '*artieri*' of the Renaissance, some of them primarily architects, that keeps together the deep structure of earth's metabolism and the phenomenological manifestation of its surface. It is distinct from the landscape notion that originates from the later Northern-European painting to inform the landscape design of the picturesque fashion.

7.2.9 Cultural Landscape

Anthropic material networks with their rural, urban, and infrastructural shells and the anthropic system of intangible fluxes of activities, energies, and information traversing the *ur-Landschaft* compose a metabolist model of landscape. As recurrently noted, a mature notion of territory lies in Ambrogio Lorenzetti's frescoes in the Palazzo Pubblico of Siena, where the *Effects of Good and Bad Government* are embedded in the topography of the anthropic hillsides surrounding the city. The frescoes of Siena date back to 1334 and depict an intangible regime imprinted on a spatial configuration. The very idea of landscape entails a cultural self-awareness that, beyond the territory and its metabolism, implies an aesthetic, intersubjective interpretation of its configuration and functioning. Be the frescoes just what we would now call a diagram (nonspatial metabolism), or a pictorial representation of the *paese* (spatial territory), or all that plus its cultural framing (cultural landscape), the conventionally posited Sienese epiphany of landscape, is certainly debatable in more than one sense (Fig. 7.2).

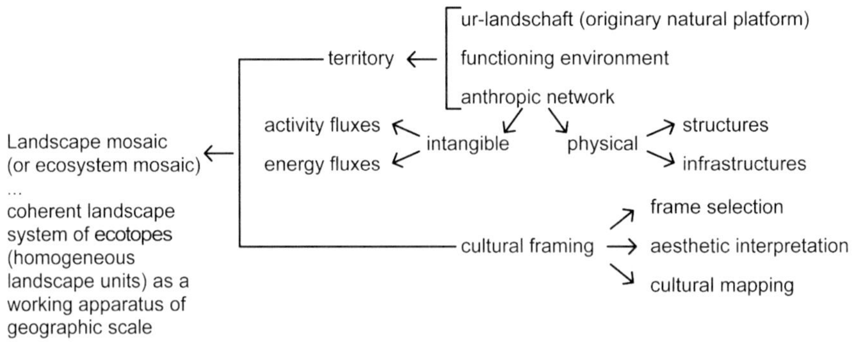

Fig. 7.2 Conventional landscape paradigm (rp)

7.3 Continuities and Discontinuities in the Landscape

7.3.1 Historical Discontinuity of the Notion of Landscape

The idea of landscape as a modern notion *tout-court* appearing in Western culture from scratch with the Italian Humanists, derives from the title chosen by Jacob Burckhardt for Subchapter III, Fourth Part, Second Volume of his monumental *The Civilization of the Renaissance in Italy*, first published in German in 1860: 'The Discovery of Beauty in the Landscape' (Burckhardt 1876, pp. 25–40). However, by that, rather than the appearance of a totally novel concept, Burckhardt is primarily determined to highlight a fracture in between antiquity and modernity, one could say in between Saint Augustine's late-antiquity, still laden with pagan legacy, and the early Italian manifestations of Humanism in the poetry of the 'wandering clerks', in the twelfth century (Burckhardt 1876, p. 27).

The term 'discovery' is, thus, marking a discontinuity in between the ancient and the modern notions of landscape, rather than a new paradigm. Certainly, "among the ancients [...] art and poetry had contained themselves within the representation of the full cycle of human life, before turning to compose the representation of nature, and, even then, the latter was always restricted within a limited ambit" (Burckhardt 1876, p. 25). Certainly, the ancient sentiment of the landscape is to Burckhardt primarily exerted through a pantheistic acknowledgement of the physical components of the surrounding nature, the spirit of a river, a cave, or a mountain. Therefore, the Tuscan landscapes described by Aenea Silvius Piccolomini in his *Commentarii* exhibit to him the shine of "modern taste, not imitation or influence of Antiquity". Aeneas Silvius "passes days of unclouded happiness" sojourning with his papal court in the Longobard monastry of San Salvatore near the summit of Monte Amiata, to escape the unbearable sultriness of the summer of 1462 not less than the plague that makes the lowlands inhabitable, and leaves seducing accounts of those delightful moments among other episodes from his life. From the Amiata, he fancies of Sardinia and Corsica, while recognizing faint profiles at great length off-shore, observes the hunt of an enormous deer chased by his dogs, and enjoys "the green glaze of the meadow, where no thorns wound the foot and no insects or snakes threaten man's life"[17] (Burckhardt 1876 , p. 36).

7.3.2 A Landscape of Antiquity

Despite the absolute modernity attributed by Burckhardt to Aeneas Silvius' landscape scenes, those passages cannot sound oblivious of Pliny the Younger's

[17]Quotation from Aenea Silvius Piccolomini's *Commentarii* in Burckhardt. Translation by the author.

epistolary descriptions of his estate of the *Laurentinum* and, even more literally, of the tender grasses covering the hillside of his estate of the *Tuscum*. We have argued in previous writings (Pasini 2014) that Pliny's description of his Laurentine villa in the epistle to Gallus (Pliny the Younger 1905a) contains a mature cultural perspective of landscape as a physical and cultural construct. Pliny writes to Gallus

> From the middle of these porticoes you pass into a bright pleasant inner court, and out of that into a handsome [triclinium] running out towards the sea-shore; so that when there is a south-west breeze, it is gently washed with the waves, which spend themselves at its base.

Pliny's image is particularly poetic and dense with meaning. The marine water spray traverses the continuum 'nature/architecture/man'. If we consider Pliny's description of the *Laurentinum* in conjunction with the larger description of the 'site' of the *Tuscum* (Pliny the Younger 1905b), architecture conforms to the interface between the natural environment and its inhabitants as well as between matter and the subject's concept of order. Pliny writes to Domitius Apollinaris

> You would be charmed by taking a view of this country from the top of one of our neighbouring mountains, and would fancy that not a real, but some imaginary landscape, painted by the most exquisite pencil, lay before you, such an harmonious variety of beautiful objects meets the eye, whichever way it turns.

It is an aesthetic view expanding over the geographic space that generates a landscape in its plenitude. Pliny's poetic images demonstrate a conscious 'cultural topography' existing in antiquity. In his letters, he extensively describes the aesthetic understanding of a mosaic produced by sociocultural forces acting on the physical platform.

7.3.3 *Recuperating the External Nature*

It is certainly unnecessary to mention how the horizon of those Italian Humanists evoked by Burckhardt is oriented by the polar star of Antiquity, which unavoidably plays a substantial role in the modern 're-discovery of beauty in the landscape'. Burckhardt's work, in fact, celebrates the bonds of the Renaissance scholars to the repositories of Antiquity, the 'common source', as much as the autonomy of a modern sentiment of beauty. Furthermore, the special role of Italians in the evolution of Western thought between the twelfth and sixteenth century is ascribed to their belonging to "a semi-ancient people", which seems the product of a miscegenation between the prospects of modernity and the vestiges of Antiquity.[18]

[18] "After all, in cosmography as in other matters, it is vain to attempt to distinguish how much is to be attributed to the study of Antiquity and how much to the special genius of the Italians. They see and treat the things of this world from an objective point of view, even before they were familiar with ancients, partly because they still are themselves a semi-ancient people and partly because their political circumstances predispose them to it; but they would not so rapidly have attained to such perfection had not the old geographers shown them the way" (Burckhardt 1876, p. 11).

Dante's "great power of observation" is the expression of an "empiricism" and a "spirit of exploration" peculiar and characteristic of that semi-ancient people (Burckhardt 1876, pp. 13–4). The spirit of observation that drives the formation of the botanical gardens, pleasure gardens, zoological collections, zoology in general and the selection of canine and equine races, and even the *menageries* sporting 'savages' of diverse 'human races' in the Italian courts, will spark the diffusion of the modern interest in the exploration of the natural sphere and, beyond agrarian utility, the art of gardening the landscape in the whole of Europe (Burckhardt 1876, pp. 17–22).

The first source of Burckhardt's analysis of the 'Discovery of Beauty in the Landscape' is the accumulation of ardent explorations of the geographic space contained in Alexander von Humboldt's second volume of *Cosmos*, published in German in 1847. According to von Humboldt, in fact, "after the destruction of the ancient world—we find in the great and inspired founder of a new world, Dante Alighieri, scattered passages which manifest the most profound sensibility to the aspect of external nature."[19]

Only to complete von Humboldt's commentaries, Burckhardt proceeds to his celebrated recount of Petrarca's purported ascent of Mont Ventoux in the month of April of 1336 and his subsequent pondering on Saint Augustine's *Confessions* (Burckhardt 876, pp. 30–1).

> And men go abroad to admire the heights of mountains, the mighty billows of the sea, the broad tides of rivers, the compass of the ocean, and the circuits of the stars, and pass themselves by (Augustine 1909, p. 229)

In Petrarca's recollection, these are the verses that move him towards what will be called the modern 'discovery of beauty in the landscape'. According to Burckhardt, Petrarca is "one of the first perfectly modern men" (Burckhardt 1876, p. 28) and, at once, one of those semi-ancient eager observers of their natural environs. Via Burckhardt, his Mont Ventoux' adventure will bridge over centuries linking the ancient precedent of Philip's ascent of Mount Emo in Macedonia, narrated by Livy, and the modern sequel of von Humboldt's scientific escalation of the Chimborazo.[20]

What in Saint Augustine sounds like a sudden momentum, a contained impetus, veiled in the moral call of Christianity for the abandonment of materiality in favor of spirituality, is a surviving fragment of the sense of sublime of Antiquity and the anticipation of its modern resurrection. That momentum will then resonate in the

[19]From *Cosmos*, Burckhardt derives the sequence of Dante's observations of nature: "the bursting of the clouds and the swelling of the rivers", "the sweet breath of morning, and the trembling light on the gently agitated distant mirror of the sea", "the pine forest near Ravenna […] where the early song of birds is heard in the tall trees" (Humboldt, Cosmos: Sketch of a Physical Description of the Universe 1848, pp. 50–51).

[20]Burckardt's quotations in this sub-chapter are translations by the author from the Italian edition in the References (Burckhardt 1876) in confrontation with the classic English version of 1878 by S. G. C. Middlemore.

unconditioned surrender of the self to the landscape for the Romantic spirits of the second half of the eighteenth century, re-appropriating the sublime by retracing, through a modern sentiment, the laws of the metabolism of nature.

7.3.4 Sublime and Metabolic Landscapes

The sublime and the metabolic sentiments will be perfectly merged into one in the magnificent representation of nature inscribed in the *Géographie des plantes équinoxiales: tableau physique des Andes et pays voisins*,[21] more than a graphic board, a "colossus of information" (Jackson 2009, p. 25). The board does not simply illustrate the *Essai sur la géographie des plantes*,[22] but rather constitutes the main part of the work, published in 1807, following Alexander von Humboldt and Aimé Bonpland's world acclaimed voyage to the equinoctial regions.

By interlacing graphic, textual, and notational information, the *tableau physique*, as Humboldt comments, "contains almost the entirety of the research I carried out during my expedition to the Tropics." The *tableau physique* is a total vision of nature that, well beyond exhibiting the continental section of the equinoctial regions, aims at "connecting together all the phenomena and productions on the surface of earth." On the one hand, the *tableau physique* represents the mechanics of the planetary metabolism: "the general equilibrium [...] is the result of an infinite number of mechanical forces and chemical attractions [and] the study of nature [...] demands the gathering together of all the knowledge dealing with modifications of matter." On the other hand, Humboldt reflects, "if my tableau were capable of suggesting unexpected analogies [...], it would also be capable of speaking to the imagination and providing the pleasure that comes from contemplating a beneficial as well as majestic nature [...] lifting us to the most sublime considerations" (Humboldt and Bonpland 2009, p. 79).

The *tableau physique* is the representation of the planetary mosaic: the relational dynamics of the geographic (horizontal) distribution of plants, rearranged over the vertical axis of the equatorial ascent of the Chimborazo. The planetary mosaic is transfigured in a composition of luxuriant pictorial descriptions, precise scientific measurements, concise empiric observations, and breathtaking lyric evocations, scientifically describing poetic phenomena such as the 'variation of the blueness of the sky'. Von Humboldt's *tableau* would literally inspire Johan Wolfgang von Goethe's *Sketch of the Main Heights of the Old and New Continents* of 1814, and, more broadly, the great spirits of numerous generations pursuing "art and science as complementary routes to understanding the world" (Jackson 2009, p. 18).

[21]*Geography of Equatorial Plants: Physical Tableau of the Andes and the Neighboring Countries.*

[22]*Essay on the Geography of Plants.*

7.3.5 Historical Continuity of the Notion of Landscape

Despite any possible intermediate discontinuities, the ancient gaze that Pliny unfolds over the shore at the *Laurentinum* and the hillsides at the *Tuscum* overtly embrace the landscape and interiorize its phenomenological and conceptual understanding. Beyond the physical layers of the interaction between inhabitant and context, Pliny's passionate descriptions contain a conscious and sophisticated cultural framing, locating the sites in the local surroundings as well as placing them in the larger geographic and sociopolitical structure. This cultural self-awareness cannot be disregarded as a byproduct of informational and social levels of interaction. By virtue of its spatial and aesthetic content, Pliny's *triclinium* of the *Laurentinum* converts into a fabulous time-capsule and casts the *facies* of the ancient landscape onto the geographic perspective of the contemporary landscape. The spatial assemblage around Pliny's *triclinium* is consistent with the spatial assemblage we described around the *triclinium* of the Pompeian House of the Golden Bracelet a few hundred miles south on the Tyrrhenian coast. The two spatial assemblages will also be linked by the geological events of the eruption of Mount Vesuvius, via another Pliny, the older uncle, and his scientific observations.

In fact, while clouds of lethal gases and pumice-stones are covering the Pompeian terraced gardens, Pliny the Elder, stationed at Misenum with the fleet of Rome at his command, launches the warships towards the ascending column of smoke, with the purpose of bringing help to residents of Stabiae, but equally enticed by his 'scientific acumen' to collect empirical observations of the phenomenon. The scientific spirit of those observations that would result fatal to him, as so emotionally narrated by his young nephew (Pliny the Younger 1905c), does not essentially differ from the spirit of von Humboldt's empirical observations on the Chimborazo, eighteen centuries later.

7.3.6 Other Continuities and a Novelty in the Notion of Landscape

John Dixon Hunt's model of the 'trio of natures' that we have discussed in Chap. 2 describes a double continuity of the landscape concurrently deployed in space as well as in time. The axial geometry of an eighteenth-century English estate, extending from the focus of the edifice through the *parterres*, gardens, vegetable gardens, orchards, intensively cultivated ager, extensive fields, open nature, to finally point at some distant natural emergence such as the untamed topography of a mountain, records the gradual unfolding of man's mastery over the landscape. We have seen how that progression of visual, utilitarian, scientific, technical, and cultural control over the surrounding space described by Hunt is kindled anew after a rupture by the pleasurable observations following the first mountain ascensions of the Humanists. In Hunt's narrative, the spatial progression of the scheme from edifice to wilderness corresponds to the historical succession sparked by the first terraced parterre of the Medici in Fiesole, through their grander Tuscan estates, the villas of Latium, the French palaces with gardens, and the English landscape master-plans with edifices, geometric gardens, and picturesque parks.

Hunt's aristocratic model of double continuity (in space and history) of the landscape is founded upon the intellectual platform of elevated Humanism and its reinterpretation of the legacy of Antiquity. Equally elaborating on Burckhardt's critical reconstruction of the Renaissance, Camporesi's landscape model relies, instead, on the material culture of the poor, the industrious experiments of metallurgic craftsmen and mining engineers, and the diaries noted by wandering friars and even charlatans, rather than on the patrician epistles lingering on the pleasures of natural and manmade sceneries. So that the sea-urchin lunch recorded by fray Alberti in a day of May of 1526, during a stopover *en route* along the Neapolitan Riviera, is a perfect counterpoint to Aenea Silvius' commentaries of the Monte Amiata. In Alberti's vivid memories, the travelers gaily descend onto the shore of the lost city of Baiae and restore themselves in the sprays of the breaking waves while sea water torrents flow through the ancient ruins under their feet (Camporesi 2016, p. 116).

7.3.7 An Aboriginal Idea of the Landscape

As discussed at the beginning of this book, Hunt embeds in the landscape model of the 'trio of nature' the idea of a continuous line of evolution starting with the transition from the hunting-gathering to the herding-cultivating condition.[23] With the appearance of the manmade space of agriculture (second nature), also wilderness itself (first nature) is born as a thinkable idea by virtue of gaining its antipode. The new sedentary communities dedicated to agriculture will ground into the land nuclei of urban space and territorial networks (further articulations of the second nature) until they achieve a condition of surplus allowing to drop utilitarian actions.[24] Resources can be destined to the implementation of the 'greater perfections' of pleasure gardens and landscapes (third nature), a fully formed landscape. Camporesi also bequeaths to us one of the most poetic recounts of the material evolution of the territorial network into a lofty construct of landscape, a literary allegory of the matter in its becoming. But landscape, in more elemental forms, can be also recognized in an aboriginal construct simultaneously born with man's awareness of the surrounding space.

[23]See Chap. 2 of this book.

[24]The idea of the equivalence of the agrarian and urban space is developed in Aldo Rossi's discourse on *The Urban Artifact as a Work of Art*. Rossi quotes Carlo Cattaneo's precursory statement of 1845 on the 'artificial' character of any inhabited places (Rossi 1982, p. 34), 'distinguished from wilderness' inasmuch as they are an 'immense repository of labor' forging 'our artificial homeland' (Cattaneo 1956).

7.4 Combining Metabolist and Culturalist Positions

7.4.1 A True Reflection of His Soul

According to Ellis, geogenic biomes, conventional types of homogeneous ecological environments originally covering the full tessellation of the planet's surface, are in the present Anthropocene largely extinct and reduced to a theoretical classification (Ellis 2015). Biomes survive at best as meager relics of ages predating the planetary expansion of man, in impervious margins. Since the appearance of agrarian societies, in fact, the majority of space on earth has been composed of environments generated by heavy anthropic impact on the natural platform.

The construct of landscape in general terms is, in fact, generated by the interaction of man with space on multiple levels, involving both metabolic and cultural spheres: physical, intangible, relational, sociological, intersubjective, informational, cultural, aesthetic, perceptive, and atmospheric, adding layers to the growth of landscape paradigms. In many ways, landscape is the changing projection of our image in space or 'the faithful image of man's soul'. In his short novel *Sturm*, Ernst Jünger describes the transforming landscape on the Western Front due to the trench warfare of World War I with the following words:

> The flora of the landscape had become strangely altered, since scythes no longer cut it. [...] A different smell, hotter and wilder, lay over the fields now. And the animal world had undergone a similar transformation. [...] Man had done all of that. In his soul a change occurred, and the landscape received a new face. Man was at the root of everything, yet [his action was so violent that he sometimes failed to recognize himself] in it. These nights in the wasteland, illuminated by flashes of lightning and dazzled by the uncertain shimmer of flares, gave a true reflection of his soul. (Jünger 2015, Sect. 7.3)

When reading these lines, the exhausted visage of Lieutenant Christian Diestl in *The Young Lions*[25] comes to mind. He is gripping on to the handlebar of a military motorbike drifting away in a desperate fugue through the North-African desert, devastated by World War II's furious bombings. Lieutenant Diestl is masterfully impersonated by Marlon Brando. The transformation of Brando's countenance throughout the movie occurs in an extraordinary correspondence with the surrounding landscape. The desert scene is specular to the opening scene of the movie prior to war's outbreak, in which the charming ski-instructor Diestl is courting a pretty American tourist during a New Year's Eve gala in a cottage amidst the forests of the Bavarian Alps. The terse hillsides covered with a soft layer of snow, the majestic conifers, the crystalline air under the light of the moon resonate in the blond, lofty, and smooth elegance of Diestl's poise. The movie offers a strong representation of the correspondence between landscape and man's soul.

[25]*The Young Lions*, USA 1958, director Edward Dmytryk, adapted from homonymous novel of 1948 by Irwin Shaw.

7.4.2 Cognitive, Semiotic, Semethic, and Cultural

The centrality of the human subject is implicit in the culturalist landscape models that we have discussed, due to the constitutive apparatuses of intersubjective projections. The adoption of perceptive and semiotic schemes of cognitive landscape models provides an alternative lens that mitigates the anthropocentric orientation. In *Ecology, Cognition and Landscape*, Almo Farina develops the 'eco-field' paradigm intended to reconcile the cultural construct of landscape with the scientific approach, in the frame of a 'semethic' formulation of landscape (Farina 2009).

'Semethic' refers to a landscape mosaic containing 'meanings' both semiotic, embedded in formal patterns, and ethic, embedded in behavioral or processual patterns (Farina 2009, pp. 128–9).[26] A semethic landscape implies interchangeability between the background scene and the organisms active on it (flora and fauna, or site and population). The 'semethic' model interchanges the 'meaning-carriers' with the 'meaning-utilizers' (Farina 2009, p. 105), marking a radical overthrowing of the anthropocentric perspective.

In keeping with the scientific-cognitive approach, Farina's 'eco-field' is based on the idea of the generation of the landscape from the perceptive relation between landscape mosaic and observer (Farina 2009, pp. 108–10). In a non-anthropocentric vision, the observers within a given landscape mosaic multiply exponentially in consideration of each species retaining a species-specific perception of that mosaic. The perception of each population and each individual within the same species also may vary, as well as each functional trait associated with each individual may modify substantially the perceived environment relative to different fundamental functions. The eco-field model generates a proliferation of simultaneously coexisting 'perceived worlds' that precipitates in Farina's definition of 'cognitive landscape'[27] (Farina 2009, pp. 112–3, 122–4). The cognitive landscape embraces the semiotic interaction of the animal world, that is the observer's interpretation of the perception, as well as the cultural interpretation of the landscape mosaic (both natural and anthropic) proper of man's conceptual constructions. The latter implies the processing of perceptions and their intersubjective sharing within a community of humans (Farina 2009, p. 137).

7.4.3 Multiplicity of Eco-Fields

Farina refers to Jakob von Uexküll's use of the term 'environment' ('*Umwelt*') to describe the multiple singularities of eco-fields as individual but interconnected spheres. Generated as separate meaning-based structures, they link 'meaning-carriers'

[26]Farina derives the concept from Jesper Hoffmeyer.

[27]About the cognitive landscape and its components, see Chap. 8 of this book.

and 'meaning-utilizers'. Conventionally, it is an 'animal subject' that generates an 'environment related to a functional trait' (Uexküll's 'functional cycle'), such as 'medium', 'nourishment', 'enemy', and 'sex' (Uexküll von 2010, pp. 144–5).

Farina uses a well-known image from Uexküll, describing the spider's web as an idealized image of the fly, to represent one of the multiple perceptive environments coexisting within the same landscape mosaic. Uexküll says that the spider is a utilizer of meaning so precisely attuned to the fly as a carrier of meaning, that the spider's web can be described as a faithful rendering of the fly. The spider, in fact, conforms "the size of its mesh [...] to the size of the fly's body", "measures the resistance of the threads it spins by the living power of the fly's body in flight", "spins the radial threads of the web tighter than the circular threads, so that the fly is enclosed upon collision by the flexible circular threads and must certainly get stuck on their sticky droplets", while the radial threads smoothly slide the spider into the prey over the shortest path (Uexküll von 2010, p. 158).

Since the spider has not seen the fly before, Uexküll induces that its web cannot represent the fly, but rather the fly's primal image ('*Urbild*'), which as a 'primal score' is encoded in the 'primal score' of the spider. The structure of each environment can be described as a plotted plan of consumption by which the meaning-utilizer dramatically reformulates the meaning-carrier (Uexküll von 2010, pp. 158–60).

If we compose the spider/fly plot with another image from Uexküll, describing four coexisting environments in a blooming meadow, the cognitive landscape composed of eco-fields acquire a vivid appearance. The construction plan of a flower stem (meaning-carrier) in that meadow is disrupted while its components are ripped apart to be inserted into completely different construction plans of a 'flower-picking girl', an 'ant', a 'cicada larva', and a 'cow' (meaning-utilizers) to end up converted into respectively 'ornament', 'path', 'spigot', and 'clump of food' (Uexküll von 2010, p. 144).

> "Anything and everything that comes under the spell of an environment" concludes Uexküll "is either redirected and re-formed until it becomes a useful carrier of meaning [while] the original components are [...] crudely torn apart without the slightest consideration for the structural plan which controlled them to that point". (Uexküll von 2010, p. 145)

As a model of landscape ecology developed on Uexküll's formulations and based on multiple eco-fields, Farina's cognitive landscape forms an ample scientific, non-anthropocentric construct, capable of encompassing the idea of cultural landscape. Besides the eco-fields, the cognitive landscape construct involves components such as NbL (Neutrality-based Landscape), IbPL (Individual-based Perceived Landscape), and IbCL (Individual-based Cognitive Landscape) depending on the multiple levels of subjectivity involved (Farina 2009, pp. 18–20) (Fig. 7.3).

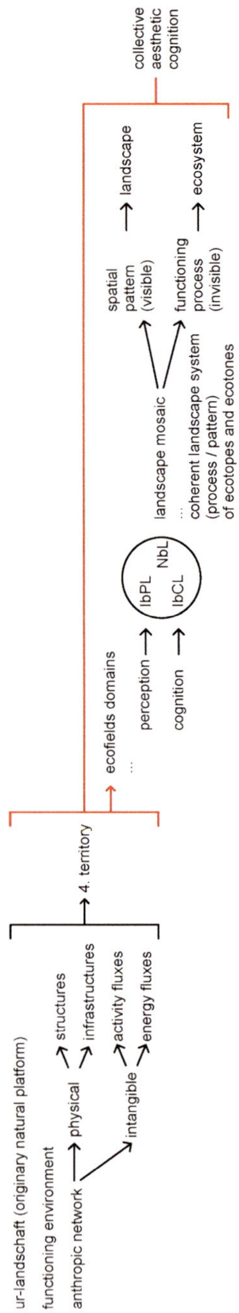

Fig. 7.3 Landscape paradigm combining scientific-cognitive and geophilosophic-atmospheric components (rp)

7.4.4 *Non-anthropocentric Mindsets*

Farina's model of the eco-fields is particularly significant to a contemporary insight on landscape as it frames cultural perspectives in a perfectly describable, non-anthropocentric paradigm neither resorting to mystic holisms nor to agnostic relativisms. The core issues of today's debate on ecological emergency are the formation of a constructive mindset capable of overcoming the rapacious anthropocentric bias that has under-arched the baleful exploitation of natural resources.

In his encyclical letter *Laudato Si'*, Pope Francis urges to care for our 'common home' (Francis 2015). By that, he refers to earth as a habitable space, where 'common' does not simply mean shared by the human population, but rather shared by the community of all the different living species and abiotic things that compose the terrene assemblage, in concordance with the Franciscan inspiration of his pontificate. Pope Francis warns explicitly against 'misguided anthropocentrism', whose consequent 'practical relativism', guiding the lifestyle of our age, sets man's immediate convenience as a priority, while all else becomes relative (Francis 2015, p. 90).

Regarding nature as a provision of ecological services and regarding subaltern individuals as a provision of labor are equivalent postures that drive to environmental degradation and social decay. Practical relativism, in fact, summons "the invisible forces of the market to regulate economy, [considering] their impact on society and nature as collateral damage" (Francis 2015, p. 91). The 'external forces' evoked to reshape the landscape-urbanist *agoras* of our days,[28] can be regarded a self-absolutory, manicured expression of this attitude.

The revision of the positioning of the human self in space is, thus, broadly involving the contemporary conscience spanning from the mondialist, ecologist, anti-globalization movements calling for a change in mindset. That means for us a new 'landscape construct', reconciling the three ecological spheres of nature, society, and mind that Guattari calls ecosophy.

7.5 Chapter's Conclusions: Multiple, Fluctuating, Simultaneous

Landscape cannot be seen as monocular, stable, or longitudinal, but rather 'multiple', as a composite aggregate of snapshots aiming from diverse viewpoints, 'fluctuating', as the shifting trajectory of a nonlinear metamorphosing construct, and 'simultaneous', as precipitating into the vertical axis its entire set of diverse layers and components, poising between the antipodal attraction of living and mechanistic principles.

---

[28]See Chap. 2 of this book.

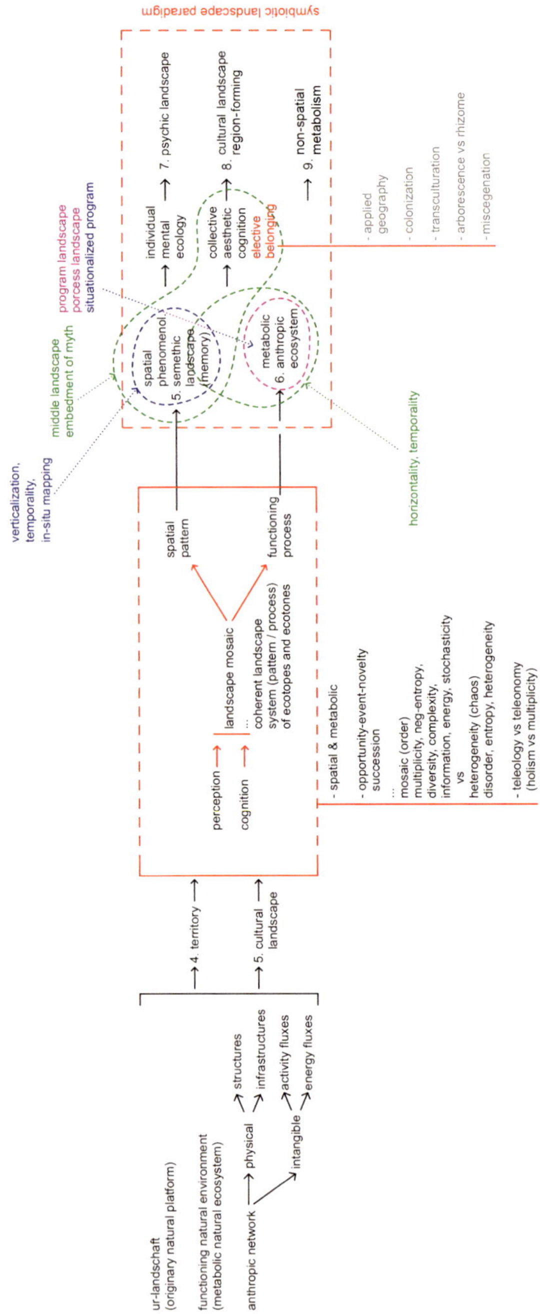

Fig. 7.4 Symbiotic landscape paradigm (rp)

The diagram that follows proposes a possible 'symbiotic landscape paradigm'. The diagram assembles the elements of a traditional construct of landscape with others, derived from contemporary branches of the interdisciplinary debates touching landscape-related themes appearing throughout this book. Its components originate from a recent or distant past. Some of them are commonly in use, others, for various reasons, have been left aside, intentionally discarded, or inadvertently disregarded in the recent debate on landscape. The typical layers of a 'culturalist-geopolitical' perspective are combined with levels typical of 'metabolist-cognitivist' visions. *Ur-Landschft*, habitat, environment, metabolic ecosystem, anthropic network, with tangible and intangible fluxes of matter, energy and people, territory, and cultural landscape, are interfaced with a cognitive pattern-and-process landscape mosaic resulting from the accumulation of multiple eco-fields. By means of engrafting a series of semiotic, geographic, philosophical metabolist, and cognitivist elaborations, the mixed paradigm becomes a ductile instrument for material and discursive analysis of contemporary space (Fig. 7.4).

References

Augustine S (1909) The confessions of saint augustine (trans: Pusey E). The Collier Press, New York [circa 400]

Böhme Gernot (2006) Atmosphere as the subject matter of architecture. In: Ursprung Philip (ed) Natural histories. Zurich, Lars Müller

Baccini P, Brunner P (2012) Metabolism of the anthroposphere. The MIT Press [1991]

Bonesio Luisa (2002) Oltre il paesaggio. Arianna Editrice, Bologna

Burckhardt J (1876) La civiltà del Secolo del Rinascimento in Italia, vol II (trans: Valbusa D). Sansoni [1860]

Camporesi P (2016) Le belle contrade. Nascita del paesaggio italiano.Il Saggiatore, Milano [1992]

Cattaneo C (1956) Industria e morale. In: Relazione alla Società di Incoraggiamento d'Arti e Mestieri, Milano 1845, Scritti Economici, vol III. Le Monnier, Milano [1840]

Chevrier J-F (2010) Programme, monument, landscape. *El Croquis* Herzog and de Meuron 2005 2010 (152–153)

Chevrier J-F, Herzog J (2010) A conversation with Jacques Herzog and Pierre de Meuron. *El Croquis* Herzog and de Meuron 2005 2010 (152–153)

Deluze G, Guattari F (1987) A thousand plateaus: capitalism and Scihzophrenia (trans: Massumi B) The University of Minnesota Press, Minneapolis [1980]

D'Angelo P (2010a) Estetica Ambientale. Enciclopedia Treccani. Istituto dell'Enciclopedia Italiana, Roma

D'Angelo P (2010b) Introduction. In: Estetica del Paesaggio. Quodlibet

Ellis E (2014a). (Anthropogenic taxonomies). A taxonomy of the human biosphere. In: Reed C, Lister N-M (eds) Projective ecologies. Actar, New York, Barcelona

Ellis E (2014b) Ecologies of the Anthropocene. Global upscaling of the socio-ecological infrastructures. In: Ibañez D, Katsikis N (eds) Grounding metabolism: new geographies, n. 6. Harvard GSD, Cambridge

Ellis E (2015) Ecology in an anthropogenic biosphere. Ecol Monogr 85(3):287–331

Farina Almo (2009) Ecology, cognition and landscape. Springer, New York

Francis (2015) Encyclical letter Laudato Si' of the holy father francis on care for our common home. Vatican Press, Vatican City

Humboldt von A (1848) Cosmos: sketch of a physical description of the universe, vol II (trans: Sabine E). Longman, Brown, Green, and Longmans/John Murray, London [1847]
Humboldt von A, Bompland A (2009) Essay on the geography of plants. In: Jackson ST (ed) (trans: Romanowski S). The University of Chicago Press, Chicago, London [1807]
Ibañez D, Nikos K (2014) Editorial. In: Grounding metabolism: new geographies, n. 6. Harvard GSD: Cambridge
Jackson ST (2009) Introduction: humboldt, ecology, and the cosmos. In: von Humboldt A, Bonpland A (eds) Essay on the geography of plants. The University of Chicago Press, Chicago, London, pp 1–46
Jünger E (2015) Sturm, tr. A. Walker, Telos Press 2015, section 7.3
Magnaghi Alberto (2010) Il progetto locale. Verso la coscienza di luogo. Torino, Bollati Boringhieri
Nietsche F (1997) Untimely meditations (trans: Hollingdale RJ). Cambridge University Press [1873–6]
Pasini R (2014) Triclini sul mare e rotte culturali. *Graphie* (Il Vicolo) (66)
Pliny the Younger (1905a) Letter XVII. To Gallus. In: Bosanquet FCT (ed) The letters of Caius Plinius Caecilius Secundus, Book 2 (trans: Melmoth W). George Bell and Sons, London, pp 62–68
Pliny the Younger (1905b) Letter VI. To Domitius Apollinaris. In: Bosanquet FCT (ed) The letters of Caius Plinius Caecilius Secundus, Book 5 (trans: Melmoth W). George Bell and Sons, London, pp 152–156
Pliny the Younger (1905c) Letter XVI. To Tacitus. In: Bosanquet FCT (ed) The letters of Caius Plinius Caecilius Secundus, Book 6 (trans: Melmoth W). George Bell and Sons, London, pp 193–198
Rossi A (1982) The architecture of the city. The MIT Press, Cambridge [1966]
Steiner H (2014) After habitat, environment. In: Grounding metabolism: new geographies 6. Harvard GSD
Uexküll von J (2010) A theory of meaning. In: A foray into the worlds of animals and humans (trans O'Neil JD) University of Minnesota Press, Minneapolis [1934, 1940]
Ursprung Philip (2006) Beauty and atmosphere. In: Ursprung Philip (ed) Natural histories. Zurich, Lars Müller

Part III
Tests, Philosophy, Agenda

Chapter 8
Tests for a Symbiotic Matorral

Abstract This chapter reports about the Tests for a Symbiotic Matorral landscape installation, implemented in the mountains of northeastern Mexico. The installation is located in a natural site of the Parque Ecológico Chipinque impacted by the infrastructures of the expanding metropolitan area of Monterrey. What follows is a characterization of the installation as an in situ (physical and spatial) transcription of the symbiotic landscape paradigm, proposed in previous chapters, combining culturalist and metabolist components. The text describes the organization of the geophilosophical (culturalist) and the cognitive (metabolist) spatial models selected as bases for the project, respectively Deluze and Guattari's 'becoming' and Farina's 'cognitive landscape' paradigms. Then, the text explains how the installation plots in space the superposition and the coincidence between the two models. The landscape installation is described both in its general structure and detailed articulations, as a natural narrative where the physical topography resonates in an under-arching mythology. The conclusions illustrate the social, educational, and scientific aims of the landscape installation and its management in the frame of the campaigns of local institutions for raising ecological awareness among the young generations.

Keywords Symbiotic Matorral · Landscape installation · In situ transcription
Deleuze Guattari · Geophilosophy · Metabolist construct · Almo Farina
Cognitive paradigm · Ecological awareness

8.1 Introduction

The research project on the built landscape that is narrated in this book, aims at the reconciliation of 'culturalist-geophilosophical' and 'cognitive-metabolic' perspectives over the contemporary continuum. To that end, we have proposed a symbiotic landscape paradigm, combining elements and layers ascribable to both positions. Beyond the theoretic elaborations, an exercise of physical transcription of the symbiotic paradigm is being pursued through the implementation and subsequent observation of a landscape installation in the Sierra Madre Oriental, described

R. Pasini, *Landscape Paradigms and Post-urban Spaces*, The Urban Book Series,
https://doi.org/10.1007/978-3-319-77887-7_8

hereby (Fig. 8.1). The installation is the in situ superimposition of two distinct landscape paradigms: one geophilosophical, Deleuze and Guattari's 'becoming' construct, and one cognitive, Almo Farina's 'cognitive landscape' construct. Eventually, the project advocates for a perception of our living space as both a functioning ecosystem and a cultural landscape.

8.2 Overlapping Constructs

8.2.1 Becoming

On the one hand, the 'becoming-' process ('becoming-animal', 'becoming-woman', 'becoming-child') is projected over the BwO's (Body without Organs) construct, multiplied throughout Plateau 10 of *A Thousand Plateaus* into a cloud of quasi-equivalent nomenclatures, such as the 'plan(e) of consistency', 'plan(e) of haecceity', 'plan(e) of consistency of Nature', and 'field of immanence', containing non-subjective individuations defined by latitude (intensity) and longitude (speed). (Deluze and Guattari 1987, pp. 256–7) With the 'plan(e) of consistency of Nature', Deleuze and Guattari represent an immense 'Abstract Machine' conformed as a surface of immanence where all multiplicities exist according to degrees of intensities and acceleration or deceleration. (Deluze and Guattari 1987, p. 254) Furthermore, the multiplicity of plan(e)s of immanence are defined in sheer, but also inseparable, juxtaposition to another multiplicity, that of the 'plan(e) of transcendence' (Deluze and Guattari 1987, p. 256), or 'structural plan(e) of formed organizations', 'genetic plan(e) of evolutionary developments', containing subjectivities and forms organized in serial or proportional structures. The nebulization of the 'plan(e)s' of immanence and transcendence responds to the authors' 'ethos', tending to replace both subjectivities and individuations with vaporous multitudes.

8.2.2 Cognitive Landscape

On the other hand, Farina's 'cognitive landscape' construct involves base components called 'eco-fields' (Farina 2009, pp. 108–9), combined into structures such as the NbL (Neutrality-based Landscape), IbPL (Individual-based Perceived Landscape), and ObL (Observer-based-Landscape) (Farina et al. 2005; Farina 2009, pp. 18–20). ObL is replaced by two terms, IbCL (Individual-based Cognitive Landscape) and SbL (Societal-based Landscape), in later formulations (Farina 2009, pp. 19, 74). Apart from slight deflections of the meanings attributed to the terms in different versions, Farina proposes an interpretation of landscape relying on semiotic mechanics. Building upon Uexküll's *Umwelt*, the 'eco-field' is defined as "the ecological space where functional traits […] intercept the resource […] according to a cognitive perception of the environment" (Farina 2009, p. 109). "If we sum all the eco-fields

Fig. 8.1 Installation site of the tests for a Symbiotic Matorral project in the Sierra Madre Oriental: luxuriant matorral with large nopal laden with prickly pears, power-line pylon, and wanderers (*photo* rp)

activated by an individual [...] the range of all possible eco-fields [is] the cognitive landscape of that species" (Farina 2009, p. 110). The comprehensive 'cognitive landscape' is, then, the superimposition of the 'eco-fields' relative to all functional traits, individuals, species, populations, communities within a given space.[1]

However, if we consider the combination of signal sources before perception, we can construe in abstract terms the mosaic of a state previous to that of landscape. Farina calls it NbL, a pattern of sources of unperceived signals, also described as the 'un-decoded landscape' or 'landscape background noise' (Farina et al. 2005, p. 236) (for the latter we must imagine a background without any foreground). NbL is in fact the comprehensive landscape mosaic available as a potential only source of sensorial perception and cognitive interpretation, therefore a permanent source of information unperturbed by any observer (Farina 2009, p. 19).[2]

A number of signals singled out and perceived by the biological sensors of an organism, instead, become a set of signs, IbPL, species-specific or even individual-specific. Signs in fact do not exist within the source, but only within the observer's perception, where signals are organized in forms, as perceived 're-sources'. A 'more profound observation' of the environment propelled by 'culture' produces ObL, that is "the anthropogenic way to perceive the surroundings" (essentially analogous to SbL) (Farina 2009, p. 237). We could say that, through culture, signs are organized in deeper forms, which exceed the identification of resources to transfigure into intersubjective values.

8.2.3 Cognitive Landscape and Becoming: A Coincidence

The 'cognitive landscape' is always depicted as the comprehensive combination of NbL, IbPL, and ObL (or its equivalents IbCL + SbL). In other terms, the 'cognitive landscape' is the sum of all the individual-specific, species-specific, trait-specific 'eco-fields', including, for humans, the cultural (that is intersubjective and social) interferences with biological senses. Within a determined space, or we could say within a field, the 'cognitive landscape' accounts for all of the 'scores' and 'melodies' (Uexküll 2010, pp. 160–1) that enact the carousel of life. In Deleuze-Guattarian terms, we could say that the 'cognitive landscape' accounts for all the 'forms' and 'subjectivities' within the serial/structural organization of the 'plan(e) of transcendence'. But simultaneously, it accounts for the whole distribution of degrees of 'intensities' (latitude) and 'speed' (longitude), that is 'non-subjective individuations', over the BwO's, 'plan(e) of immanence', immense

[1]For a detailed definition of eco-field, see: Farina (2009, pp. 108–12). For a specific method of combined evaluation of trait-specific eco-fields, see sub-chapter 'Scoring the Cognitive Landscape' (Farina 2009, pp. 112–4).

[2]The NbL is anyway another example of a 'ladder of Rome' (i.e., 'Wittgenstein's ladder'), as Farina notes that the landscape is the result of an observer's perception and, in the case of the human being, a perception driven by culture (Farina 2009, p. 74).

'Abstract Machine', or 'plan(e) of consistency of Nature'. Deleuze and Guattari in fact say that "the plane of consistency cuts across [all] multiplicities [...]. The plane of consistency is the intersection of all concrete forms" (Deluze and Guattari 1987, p. 251). That intersection converts forms and subjectivities into latitudes and longitudes, i.e., non-subjective individuations, plotting the bi-univocal correspondence between the separate spheres of immanence and transcendence.

8.2.4 NbL to Plane of Immanence as IbPL + ObL to Plane of Transcendence

If the un-decoded signals of NbL are the equivalent of the non-subjective individuations of the 'plan(e) of consistency of Nature', the organized signs of the IbPL and ObL (or IbCL + SbL) are the equivalent of the forms and subjectivities of the 'plan(e) of transcendence'. As Deleuze and Guattari's 'plane of consistency is the intersection of all concrete forms' that converts them into the Abstract Machine, Farina's NbL is the abstract section through the multitude of 'eco-fields' that removes all the interfering observers to return an un-decoded landscape, or perfect background noise.

8.3 Tests for a Symbiotic Matorral: A Landscape Installation[3]

8.3.1 Reasons for a Landscape Installation

The Tests for a Symbiotic Matorral project is a linear apparatus of landscape installations implemented in the Sierra Madre Oriental of Mexico. The project intervenes in a site of highly aesthetic value of the Parque Ecológico Chipinque, where the expanding metropolitan area of Monterrey makes contact with the pristine ecosystem of the *matorral submontano*.

The *matorral submontano* is a low, dense ecosystem covering the lower part of the mountainsides in northern Mexico. Its flora is thicket and scrubland with a varied fauna, presenting an extremely high biodiversity. A large share of the public opinion in the city still considers the *matorral* nothing but wasteland to be cleared for future constructions. The path pursues the aim of promoting awareness about the beauty

[3]The Tests for a Symbiotic Matorral is a landscape installation, result of a research project on the landscape. Research team: Roberto Pasini (principal researcher), Patricio Garza, Andrea Ramos, Fernanda Rosas, et al.; design of installation: Pasini, Garza, Ramos, Rosas; stone carving and on-site implementation: Rodrigo Legorreta; metal works: Jaime Islas; metal engravings: Jopamec; support and funds: Universidad de Monterrey, DIECI, Parque Ecológico Chipinque.

and fragility of this ecosystem among young generations. The path offers the school children of Monterrey a route equipped with informative signposting, illustrating flora, fauna, and their ecological metabolism. Specific installations along the path let the children hear like a rabbit, smell like a bear, see like a coyote, and fly like a green jay, for them to understand that humans are not the only inhabitants of the planet. Simple surveys, administered before and after the trek, monitor the change in the perception of the ecosystem and evaluate the increment in ecologic awareness.

8.3.2 Site

The installation site is set in a point in which the expanding infrastructures of the 5-million metropolitan area of Monterrey clash against the dense cover of *matorral* and its extraordinary biodiversity. The installation converts a 500-m-long segment of an abandoned trail for the maintenance of a high-voltage line into a floro-faunistic path through the luxuriant ecosystem. The access point to the intervention is marked by the presence of an imposing truss pylon of the power-line.

8.3.3 Overthrowing Perspectives

The project converts the trail into a floro-faunistic observation route that overthrows the conventional anthropocentric perspective. The route is in fact a linear apparatus composed of a series of totally reversible installations in steel, wood, and rock, that favor the observation of the ecosystem from varying vantage-points of different species populating it. The multifocal vision that the project conjures up focuses on 'deep forms' of the ecosystem, in analogy with Farina's construct of the 'cognitive landscape', that is 'spatial formations carriers of meaning', where 'functional trait' meets 'resource' and 'its cognitive perception' enacted by 'spatial configuration'.[4] The installation combines in fact points of observation and sensorial perception with functions and fundamental acts relative to diverse species.

8.3.4 A Route in Three Sections

The trail was originally traced by bulldozing a secondary topographic ridge. The operations removed a superficial layer of soil, roughly flattening the rock bottom and preventing the vegetation from obstructing the way on a section of about 2 m in

[4]For a definition of eco-field as a 'spatial configuration carrier of meaning', where the roles of 'builders' and 'users' (traditionally, plants and animals) are interchangeable, see: Farina (2009, p. 109).

Fig. 8.2 The route along the ridge of the Sierra Madre Oriental (*photo* Lgo Rodrigo Legorreta)

width (Fig. 8.2). The regrowth of spontaneous vegetation has long reoccupied the passage, leaving only a narrow width of about 60 cm, where the rock surfaces from the sediments accumulating on the sides.

The floro-faunistic route is subdivided in three sections: perceptive, experiential, and informative. Respectively called the 'fauna-trails path', the 'glade of becoming', and the 'flora path', the three sections are contained in between the 'entrance portal' and the terminal 'contemplation circle'.

8.3.5 Portal to Other Worlds

The entrance is through the first installation: a light metal structure in the shape of a portal composed by a horizontal platform lifted 40 cm above ground and surmounted by a slender square frame erected vertically. A squared rock of local limestone provides a step stone to reach the platform, decked with a grid of thin metal bars in a parallel arrangement. The left half of the vertical square frame is occupied by a metal panel bearing engraved a topographic map of the area. The map carries basic geographic information necessary for orientation, identifying the three sections of the route and the fundamental components of both installation and context. Further information about the elements of equipped paths and environmental context is provided by texts engraved on a metal band anchored on the deck of the platform, marking the actual threshold of the entrance to the route at the foot of the squared frame. The threshold materializes the gate between the sphere of

Fig. 8.3 Portal to the floro-faunistic route, detail (*photo* rp)

ordinary spatial perception and the immersion into a novel ambient, where the anthropocentric perspective will be overthrown and a different exploration of the environment is favored (Fig. 8.3).

8.3.6 Path of Fauna Trails

The first section of the route is a long itinerary of about 400 m through the *matorral submontano* with an inclination of about 15° west, on the north–south axis. It is called 'path of fauna trails' from the transversal animal trails crossing the path perpendicularly. The animal trails are identified by wooden stakes marking the tunnels in the dense foliage produced by the crossing of fauna. Metal silhouettes applied to the stakes identify the species compatible with the proportions of the passages and provide ethologic information.

8.3.7 Glade of Becoming

At the end of the first section, the branches of the bushes from the two sides of the path are interwoven to form a vegetal tunnel framing a human-sized passage (Fig. 8.4). A metal arch of the same anthropomorphic dimensions at the end of the

Fig. 8.4 Gallery of the Interwoven branches installation (*photo* Lgo Rodrigo Legorreta)

vegetal tunnel marks the entrance into a glade. The glade is an ovoid area of about 10 m across by 30 m in length, smoothly bending to the right of the visitor rotating the axis of the route by about 45° to the west. The glade represents the central section of the installation.

A longitudinal track across the area traces a wavy trajectory that follows the pristine topographic ridge. On the opposite side of the glade a wooden stake of human height marks the exit point southwest. The metal arch and the wooden stake are the two terminals, entrance and exit, of the trajectory of exploration offered to the visitor. The longitudinal track through the glade is crossed by three transversal animal trails called 'becoming-animals'. As a sequence, they acquire a character of longitudinal multiplicity that we will describe in detail later as a 'becoming-demon'.

The matrix of intersected axes of the 'glade of becoming' captures the carousel of Nature in the *matorral*. The matrix organizes the dynamics of the 'plan(e) of transcendence' over which animalities and subjectivities slide, as well as the dynamics of the 'plan(e) of immanence', the abstract section of all those forms that are converted into degrees of intensities and speeds, 'latitudes', and 'longitudes'. Over the 'glade of becoming' the two plan(e)s of transcendence and immanence reflect one another.

8.4 Deleuze-Guattarian Ascendancies

8.4.1 Oedipal, Mythic, and 'of-a-Third-Type' Animals

The visitor's interaction with the animal sphere is organized on Deleuze and Guattari's anti-taxonomic subdivision of the animals into the three kinds of 'domestic', 'mythic', and 'demonic'. The 'domestic' animals or 'anti-animals', like the 'family pets', are 'Oedipal animals' that invite humans to regress into a 'narcissistic contemplation' that we could define anthropocentric. This first kind is banned beyond the entrance portal. Of the 'mythic' kind are animals with 'characteristics and attributes' described 'in the great divine myths'. In those ancient myths, they are treated "in such a way as to extract from them series or structures, archetypes or models". The 'demonic' animals are creatures 'of a third type' related to multiplicity, whose determination we will discuss further on (Deluze and Guattari 1987, pp. 240–1).

8.4.2 Analogies by Resemblance and Proportion

According to Deleuze and Guattari, Carl Gustave Jung's, and Claude Lévi-Strauss' works is aligned with respectively the 'serial and structural analogies' that natural history deploys in its classification of the world. The 'serial analogy' is governed by the principle of 'resemblance' that requires 'imagination' and leads to the Linnaean taxonomy, while the 'structural analogy' is governed by the principle of 'proportion' and requires 'understanding' leading to Cuvierian compared anatomy. Both through the '*a* resembles *b*' and the '*a* is to *b* what a^1 is to b^1' type of analogy, nature is conceived as an 'enormous *mimesis*'.

According to Deluze Guattari, Jung's theory of the 'archetypes' uses serial analogies, made of metamorphic progressions and regressions from one term to another, to interpret the collective unconscious through the dream. In those metamorphic series, however, 'man is no longer the eminent term', but the eminent term is an animal, vegetal, or mineral, 'in relation to a given act or function' and 'in accordance with a given demand of the unconscious'.

On the one hand, Deleuze and Guattari refer their 'mythic animal' notion to the Jungian theory of the 'archetype' as 'collective unconscious', aiming at overthrowing the anthropocentric vision of the world by means of a 'serial organization of imaginary' (nature-culture-nature) (Deluze and Guattari 1987, pp. 235–6). On the other hand, Deleuze and Guattari regard Lévi-Strauss' structuralism as an attempt to overcome the serial organization of the world with a symbolic and structural order of understanding, thus based on proportional analogies, that is correspondences of relationships (Deluze and Guattari 1987, p. 237).

In particular Lévi-Strauss' theory of 'totemism' replaces Jung's animalist pro-/regression series of 'nature-culture-nature', with an analogy of proportionality through which the characters of the totemic animal are not acquired by resemblance, but rather introjected by means of proportional relationships. The 'archetype of the sacrifice', where off-spring kills father feeding on his flesh to acquire his attributes, organizes the world by metamorphoses, that is analogy of resemblance (Deluze and Guattari 1987, pp. 235–6). Conceptual metaphors replace imaginative metamorphoses. The 'institution of the totem' marks, therefore, the symbolic structuration of the world through metaphor (nature–culture relational continuity), where the attributes are acquired by relational analogy: totem replaces sacrifice.

8.4.3 *Sacrifices, Totems, and the Myth*

In Deluze and Guattari's formulation, Jung's man says -"I am a wolf", opening up a serial theme made of collective archetypes that can be recurred back and forth by virtue of metamorphic equations. Lévi-Strauss' man says -"I am to another man what the wolf is to the sheep" structuring an understanding by virtue of the 'institution of the totem' (Deluze and Guattari 1987, p. 237). One should then conclude that 'mythic animals' acquire their attributes either in a Jungian or Lévi-Straussian fashion, that is by serial or proportional analogy, and therefore myth is a narrative inhabiting a terrain in between the archetypal sacrifice and the institution of the totem.[5]

8.4.4 *Demons and Their Symbiotic Alliances*

According to Deleuze and Guattari, to the third 'demonic' kind belong animals of pack, school, band, whose identity is tight to a non-arborescent multiplicity, non-filiative population, propagation by epidemic and contagion. 'Demonic animals' create non-familial assemblages capable of symbiotic alliances beyond the taxonomic boundaries (Deluze and Guattari 1987, p. 241), such as that of the 'wasp and orchid'.

The 'wasp and orchid' imitation, where 'the orchid forms a tracing of the wasp' in a parallelism between a 'plant organization' and an 'animal organization' (Deluze and Guattari 1987, p. 10), could be described as an analogy. But it is simultaneously a 'capture of code' in which 'the orchid forms a map of the wasp' (Deluze and Guattari 1987, p. 12). Such a 'non-filiative', 'non-hereditary' metamorphosis supports a 'neoevolutionist' perspective (Deluze and Guattari 1987, pp. 238–9). While defining the 'becoming-', Deleuze and Guattri describe a 'shared deterritorialization' of the wasp becoming a "liberated piece of the orchid's

[5]"Does it not seem that alongside the two models, sacrifice and series, totem institution and structure, there is still room for something else, something more secret, more subterranean: *the sorcerer* and becomings (expressed in tales instead of myths and rites)?", (Deluze and Guattari 1987, p. 237). Tales, myths, rites are located between sacrifice and totem.

reproductive system" and of the orchid becoming "the object of an orgasm in the wasp, also liberated from its own reproduction" (Deluze and Guattari 1987, pp. 293–4).[6]

However, the multiplicity of the pack, school, or band is no obsolete social form or anyway inferior to that of family and state, but rather it represents a deterritorializing alternative. The multiplicity of the pack establishes an equilibrium with the individuality through the presence of the pack-leader and borderline members (such as the 'sorcerer' or the 'shaman'), which produces what Guattari calls 'social ecosophy' (Guattari 2000, p. 34), or processes of 'continuous resingularization' (Guattari 2000, p. 69) that cultivate anarchist *dissensus* to arborescent organization. In Guattari's vision, social ecosophy of the pack intends to replace IWC's organization of 'imaginary crowd aggregates' (serial subjectivity of salaried, mass of the uninsured, and élite of executives) with autopoietic 'collective assemblages of enunciation' capable to constantly redefine their own singular identity (Guattari 2000, p. 61).

8.4.5 Human Alliances with Demons

The 'demon', as the borderline of the pack, is the central term of the process of 'becoming-' broken down by Deleuze and Guattari into four points (Deluze and Guattari 1987, p. 61): 'alliance with a demon'; 'human being's passage by contagion into animal pack where demon is borderline'; 'second alliance with human group'; 'human being (as borderline between animal pack and human group) guides the contagion of the human-animal becoming' (that is 'symbiosis between heterogeneous terms') (Deluze and Guattari 1987, p. 249). As an alternative to demonic animals such as werewolves and vampires, Deleuze and Guattari describe the demon-shaman Don Juan in Carlos Castaneda's accounts. Don Juan's guides Castaneda through the proportional-relational becoming-animal and becoming-molecular characterized by the distinct micro-perception of the environmental elements such as water and air (Deluze and Guattari 1987, p. 249).

---

[6]Deleuze and Guattari also parallel the 'wasp and orchid' relation to Uexküll's 'spider and fly' example of 'transcoding', where the spider 'has a fly in its head', or better the fly's code, that allows it to plot its net in perfect correspondence to the fly's traits (see previous Chap. 7 of this book). The transcoding of the 'wasp and orchid', reciprocal as in Uexküll's description of 'Nature as music' with melodies in counterpoint, each serving as a motif for the other, kindles a simultaneous, bi-univocal process similar to the deterritorialization/reterritorialization dynamic. (Deluze and Guattari 1987, p. 314) A demonic symbiosis of the 'wasp and orchid' type is stunningly represented in Carol Reed's 1949 movie The Third Man, where Henry Lime and Anna Schmidt (interpreted by Orson Welles and Alida Valli) form an unbreakable consortium through war, famine, epidemic, crime, vampirehood, remaining indifferent to the moral principles of a society struggling to redeem itself from the tragic fate of WWII.

8.5 The Glade of Becoming

8.5.1 Three 'Becoming-Animals' and One 'Becoming-Demon'

We will now resume the narration of the landscape installation interrupted at the glade of becoming. The longitudinal track through the glade is intended as the trajectory of the visitors' experience of the borderline guiding their 'becoming-' and therefore called 'becoming-demon'. Three transversal installations are aligned with selected animal trails present in the site. Each animal trail connects a tunnel in the vegetation on one side of the glade with a corresponding tunnel in the vegetation on the other side. The tunnels are marked by a metal arch and a wooden stake on opposite sides. Both arch and stake are dimensioned according to the size of the animal species compatible with the tunnel. Ramps of varying depth are excavated in the soil next to each stake, allowing the visitor to reach at the point of observation of each specific animal species, namely hare, coyote, and bear.

The series of transversal installations, called the 'becoming-animals', constitute a sequential apparatus of alliance with the 'demon-animal'. The exploration of each transversal installation favors, in fact, by metamorphic analogy specific 'becoming-animals', that is the acquisition of the perceptive attributes of a mythic animal. Man belly-crawls into the first ramp, lies prone the face close to the ground, scans the tiny leaves of the bush, their vibrations maul his/her ears through the hot air. Hear like a 'hare'. Then man scrambles along the second ramp, squats, sees the preys moving, the rapid quivers of their nerves under the delicate fur. See like a 'coyote'. Man finally walks down the third ramp, his/her eye level is that of the bear. Man sees the purple prickly pears of the *nopales*, his/her nostrils smell their sweet scent perturbing the air. Smell like a 'bear'.

The three 'becoming-animals' can also be regarded as non-anthropocentric 'eco-fields', simulating the sensorial perception of other species in an unusual assemblage, hear like a hare, see like a coyote, smell like a bear. 'Becoming-animals' overlap with 'eco-fields' (Fig. 8.5).

8.5.2 Animality

The sequential experience through 'becoming-animal' installations along the glade favors an alliance, by metamorphic analogy, of the human visitor with the sensorial attributes of multiple mythic animals (hare, coyote, bear). Across the longitudinal sequence of multiple transversal installations, man's alliance is, thus, not stricken with one mythic animal, but rather with a polymorphic animality. That polymorphic animality is the 'demon' that deterritorializes man from the conventional regime of societal interchange to which he/she is anchored. The alliance with the 'demon' is kindled, man is admitted into the pack. However, man poises over the borderline for

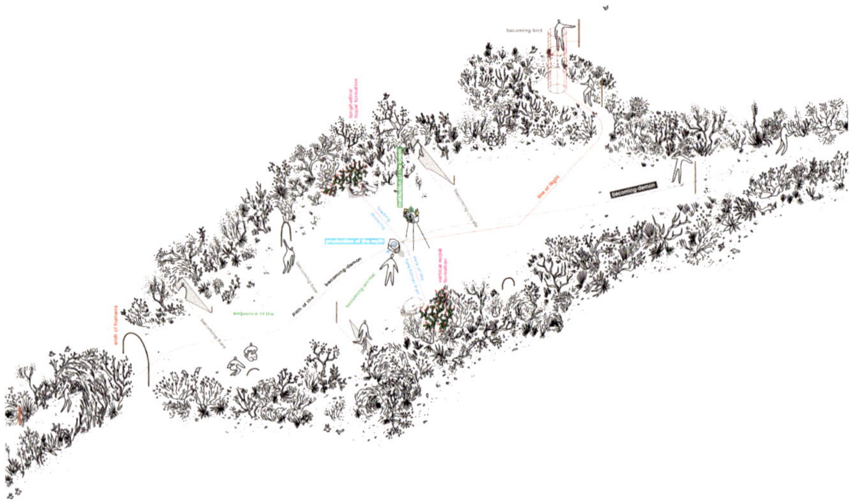

Fig. 8.5 Axonometric view of the Glade of Becoming from the Tests for a Symbiotic Matorral project. *Design* Pasini, Garza, Ramos, Rosas; *diagram* rp; *drawing* Andrea Ramos Gándara

his/her anomalous anthropic nature. As the visitors are arranged in small groups with no familial relations, the 'second alliance with a human group' proposed by Deleuze and Guattari is also attained.

8.5.3 Functional Traits and Fundamental Acts

The sequence of the 'becoming-animal' installations unfolds along the 'becoming-demon' track, reaching its climax in its median point, where the track is crossed by the transversal axis called the 'functional-trait', identifying the spot of the 'institution of the totem'.

The 'functional-trait' axis is represented by a track connecting two heterogeneous assemblages on the two opposite sides of the glade. Two large *nopales* carrying purple prickly pears are mapped roughly half way the glade: an elongated lower espalier of paddles loaded with bright mature fruits on the southeast side, corresponding to a taller but less extended formation on the northwest side with a more limited production. A flat carved rock resting on short metal feet is placed next to the southeast *nopal*, while a taller stele is placed directly on the ground next to the north-west *nopal*. The former assemblage can be described as: 'medium/large size animals come feed on prickly pears resting onto lifted rock, while small size animals underneath it'. The latter assemblage is: 'a significantly smaller number of animals come feed on prickly pears leaning to rock'. For a multiplicity of species, the feeding functional trait is accomplished at both ends, one corresponding to an abundant-resources signal, the other to a less-competed-for-resources signal. In the case of the human visitor caught in his/her process of 'becoming-', the axis of the 'functional trait' of feeding overlaps with the axis of the 'fundamental act' of eating.

The assemblage 'functional-trait/resource-perception/within-a-field' perpendicularly intersects with 'becoming-demon' to 'institute the totem'. Which also opens the issues of how intense (capacity, or latitude) (Deluze and Guattari 1987, p. 260) the flavor of the prickly pear in the mouth of the bear is tasting and how fast (speed, or longitude) (Deluze and Guattari 1987, p. 260) man is moving along the 'becoming-demon' axis (Figs. 8.6 and 8.7).

Fig. 8.6 View of the glade with horizontal nopal formation, Institution of the Totem installation, and line of flight tower (*photo* Lgo Rodrigo Legorreta)

Fig. 8.7 Snapshots from the fauna-monitoring camera documenting the crossing and feeding of wild animals over the glade of becoming: a wild boar roaming at dusk (left) and a puma hunting in the deepest hours of the night (right) (*photos* Lgo Rodrigo Legorreta)

8.5.4 Institution of the Totem

At the climax of the overall 'becoming' apparatus orchestrated over the glade, coinciding with the crossing between the transversal 'functional-trait of feeding' axis and the longitudinal 'becoming-demon' track, a rock is lifted on three steel legs in a zoomorphic assemblage. Carved with a diverse pattern of cavities varying in size, the rock supports over its uneven surface a small vegetation of *rupiculae* plants and cactuses attracting communities of insects and aviary fauna. At the middle point of the functional-trait axis, the suspended rock with its small pensile flora and fauna embodies the 'institution of the totem' with reference to the mentioned Lévi-Straussian construct. It, in fact, represents a proportional/relational *entendement* of the world alternative to Jung's animal-based metamorphic series of archetypes. A second rock lies on the ground at a short distance. The second rock is flatter and uncarved. A shallow concavity on the surface allows for water to collect in occasions of rains. As a sequence, the three transversal 'becoming-animal' axes can be considered a longitudinal assemblage. At the very crossing between this longitudinal 'becoming-animal' sequence and the transversal axis of the totem (functional trait/fundamental act), the lifted rock casts its mobile shadow onto the ground. The shadow variously intercepts the irregular profile of the ground stone and overlaps onto the ground stone's shadow to create further shades of darkness, changing reflections over the surface of the water and generating refractions into its depth.

8.5.5 Myth and Shadows Interplay

The mobile interplay of the lifted rock's shadow engaging the ground rock, its shadow, the degrees of darkness and transparency, as well as the reflections over and refractions through water, conjure up a natural narration that constitutes the very core of the myth. The interplay, as the myth, is suspended midway in between 'archetype' and 'totem', 'series' and 'structure', 'imagination' and 'understanding', 'resemblance' and 'relation'. The myth of the shadow, as a natural narration, lies at the very center of the 'becoming-demon' journey.

8.5.6 Line of Flight

At the mythic climax of the 'becoming-demon' journey man has deterritorialized his/her anthropocentric perspective and arborescent sociality. From a central position in the glade, the visitor is aligned with two gashes in the surrounding curtain of the foliage, which frame a view north to Cerro El Mirador and a view south to Pico Lobos. The views reorient the visitor reconnecting his/her experience to the geographic scale (Fig. 8.8).

There starts a deviant track, called the 'line-of-flight', that spins diagonally towards southeast to then circumvent a denser clump of shrubs, reaching a tiny

Fig. 8.8 View from the glade south towards Pico Lobos: the changing weather conditions alternate sunlight, nimbus, and mist over foliage (*photo* courtesy of Patricio Garza)

hidden clearing where a light, vertical structure is anchored. The structure is an airy truss of slender metal profiles that boxes a 3-m-high cylindrical volume sitting on its 1-m-diameter circular base. Three vertically shifted circular profiles define the perimeter of the cylinder at the base, at an intermediate height of 2 m, and 1 m above that, at the 3-m-high top. The circular profiles are vertically connected by eight slender struts. The intermediate circular profile supports a light deck of thin metal bars plotted parallel to one another. A small arched opening on a side of the deck allows for a single visitor to climb up onto it using a ladder of metal bars spanning between two vertical struts, while the top perimeter profile works as a bannister.

8.5.7 *Avian Flight Lines and Melodies*

The 'line of flight' structure is a small observatory that overlooks the dense foliage of the *matorral* extending like the soft layer of a blanket on the varied topography. The deck offers a convenient vantage point to observe landscape dynamics both internal within the *matorral* eco-tope and of interchange with the surrounding environmental patches through eco-tone boundaries. The structure is a 'rhizome'[7]

[7]"A fiber strung across borderlines constitutes a line of flight or of deterritorialization. It is evident that the Anomalous, the Outsider, has several functions: not only does it border each multiplicity,

Fig. 8.9 Line of flight installation with school children overlooking the matorral (*photo* Lgo Rodrigo Legorreta)

that sets man amidst the weft of flight lines of the *matorral* avian fauna, contained within a height of 5 m from ground. Swathed in the cloud of avian flight lines and melody scores, man is immersed in a space made of a soothing stuff that heals the trauma of his/her deterritorializing voyage (Fig. 8.9).

8.5.8 *Down the 'Flora Path'*

Down from the structure, man exits the glade to start the third segment of the journey, the 'flora path', a shorter itinerary immersed in a denser and higher layer of vegetation. As the initial segment of the itinerary followed a constant, ascension towards the flat area of the 'glade of becoming', the final segment falls into a rather abrupt descent. Along the 'flora path', signposting bears scientific information about the ecosystem singling out exemplars of characteristic species of the *matorral sub-montano*, among which a large blooming agave, a very old *nopal*, a young one, carpets of slippery moss, *encinos* etc.

[…] not only is it the precondition for the alliance necessary to becoming, but it also carries the transformations of becoming or crossings of multiplicities always farther down the line of flight" (Deluze and Guattari 1987, p. 249).

8.6 Chapter's Conclusions

8.6.1 Contemplation, Change, and the Waldgang

The immersion into the floral sphere of the *matorral* ends up hitting a steep cliff. There, the route bifurcates spinning on both sides at the foot of the relief. Along with the route, the walls of the *matorral* foliage also diverge, cut through by the topography, giving ground to a flat triangular area occupied by medium and high grasses. The triangular landing opens a transition zone in the ecosystem, the eco-tone mediating between *matorral sub-montano* and *bosque*, the mountain woods.

A 3-m-diameter circular clearing mowed in the grasses at the center of the landing, called the 'circle of contemplation', represents the lightest installation of the project and its termination. Standing in the circle, man faces dramatic discontinuities of topography and vegetation, while the perceptions accumulated along the voyage through the *matorral* sediment in his/her body. Interior upheavals resonate in the changes of the environment (Fig. 8.10).

Unobstructed views suddenly appear on multiple directions. The route splits into multiple possible trajectories. By making his/her way back, man can reverse the

Fig. 8.10 Circle of contemplation installation (*photo* Lgo Rodrigo Legorreta)

metamorphosis and rejoin human society. By abandoning the circle westwards, man reaches deep areas of the Parque Ecológico Chipinque where the route disappears and one gets lost. If man heads eastwards, he/she can follow a trail that reaches the mountain woods: it is Jünger's *Waldgang* (Fig. 8.11).

Fig. 8.11 The *Waldgang* (*photo* Lgo Rodrigo Legorreta)

8.6.2 Testing the Paradigm and Recording Response

The Symbiotic Matorral installation is meant to raise awareness about the disregarded richness of the *matorral submontano* and its exceptional biodiversity among the population of Monterrey. Part of campaigns for the promotion of ecological values and virtuous environmental practices directed to young generations, the path is specially intended to receive students from the public and private schools of the city, spanning from primary to university level. Under the supervision of guides, visitors are allowed onto the route in small groups and left free to explore the installations rather freely.

The overall landscape installation is conceived as an in situ transcription of the symbiotic landscape paradigm. Further phases of the research will monitor the response from the visitors through the recording of their impressions, commentaries, and feedbacks, along with the general reception of the initiative in the local media. The elaboration will draw evaluations on the project's impact on the community's consideration for the ecological and aesthetic values of the landscape and test the practicability of elemental symbiotic model guidelines.

References

Deluze G, Guattari, F (1987) A thousand plateaus: Capitalism and schizophrenia (trans: Massumi B). The University of Minnesota Press, Minneapolis [1980]

Farina A (2009) Ecology, cognition and landscape. Springer, New York [2006]

Farina A, Bogaert J, Schipani I (2005) Cognitive landscape and information: new perspectives to investigate the ecological complexity. BioSystems 79:235–240

Guattari F (2000) The three ecologies (trans: Suton P, Pindar I). The Athlone Press, London [1989]

von Uexküll J (2010) A theory of meaning. In: A foray into the worlds of animals and humans (trans: O'Neil JD). University of Minnesota Press, Minneapolis [1934, 1940]

Chapter 9
Conclusions: Life and Death of '14 Strength' and Agenda for the Next Landscape

Abstract This chapter collects the conclusions of the book and traces a brief agenda for future research work towards an *entendement* of the contemporary continuum, referred to as the Next Landscape. The text illustrates three main objectives marked in the agenda: the formation of a symbiotic vision, contaminating the technocratic regime governing today's physical and intellectual production with a non-positivistic, non-systematic exploration; the determination of a set of novel attributes to analyze, evaluate, and compare the contemporary condition of the expanded citizenship and its varying contexts; the acknowledgement and implementation of a novel platform for the democratic interaction in the 'post-urban cityness'; the under-arching of the spatial platform with a reformulated territorial mythology, to endow the expanded dwelling space with a new sense. The text deals with the notions of residence, monuments, urban and territorial grid, and comes to the idea of a network of interlaced 'neo-monuments'. A few final considerations of more general, philosophical nature close the chapter and open perspectives for future work.

Keywords Agenda · Next landscape · Novel urbanity · Post-urban cityness Soconusco · Tokyo · Montreal · Marseille · Pseudo-urban · Neo-monuments

9.1 Introduction

Coming to the conclusion of this book, the following notes present a basic agenda for what we could call the Next Landscape in the post-urban time. The challenges ahead of us deal with: (1) mitigating the imperium of technique with a symbiotic *entendement* of the surrounding space, unfolding in a non-positivistic and non-anthropocentric vision; (2) determining the attributes of the novel urbanity, or 'post-urban cityness,[1]' expanded over the geographic scale in extremely varying conditions; (3) implementing a novel physical platform for democratic interaction,

[1]The term 'cityness' was coined by Sudjic (2007): see subchapter 'Cityness' in Chap. 2 of this book.

R. Pasini, *Landscape Paradigms and Post-urban Spaces*, The Urban Book Series,
https://doi.org/10.1007/978-3-319-77887-7_9

that is the new *agoras* of the post-urban city-*civitas*; which corresponds to the simultaneous replotting of a novel mythology under-arching that territorial platform and endowing it with a new sense.

9.2 Contaminating the Technological Structure with a Symbiotic Entendement

9.2.1 Life and Death of '14 Strength'

In 'The Man Born of a Tree: A Mixtec Origin Myth', Thomas Ibach records the Mixtec myth of '14 Strength', the man born from a tree, dictated to him in 1976 by Sr. Serapio Martínez Ramos, a 55-year old monolingual native and resident of Santa Cruz Mixtepec in the region of Oaxaca.

> Fourteen Strength went to the cave of San Lucas so that he could chase the stones from the cave with a whip; because in the old times, the stones were like domesticated animals […]. And then when he arrived at the 'cross of the avocado tree' with the stones, the sun arose and it killed him because until then there wasn't a sun […]. When he died there, the stones also died. (Ibach 1980, p. 247)

The literal translation of the oral account of the myth manifests in all the liquid character of Mixtec thinking, especially in comparison with the free translation supplied by Ibach. Evident are essential analogies with the liquidity of the Mesoamerican idea of open space as a worldly transcription of the cosmogenic 'primordial sea' and its myth.

Free translation of the account:

> 1. With my humble pardon, I will give you some words about what happened very long ago, what the ancestors have said. 2. A man went to the mountains, he was going to the mountains. 3. He had been there for eight days when he saw the sacred tree there called tree madroño. 4. He went to it and made a hole in its side; 5. and he had intercourse with it. 6. After three or four months past he went and he saw that the tree was swollen. 7. And then he knew that the stomach of the tree was swollen; 8. the stomach of the tree was swollen, and he counted the months. 9. When the months were complete he went there; 10. and he made a hole in the tree's stomach and he saw a little man inside, it was a little man inside there. 11. And then he took that little man and he carried him home. 12. And when he arrived at his house the little man became alive; 13. and his name was "Fourteen Strengths". (Ibach 1980, p. 246)

Literal translation of the account:

> 1. Large become inside pluralizer you, going to give word when existed time old, say dead-ones old giving them word. 2. Walked one man within mountains walked he, going mountains, going he. 3. Left eight days going he mountains, and knowing he sacred within one tree called-it wood-it madroño. 4. There went he, made-hole he rib-cage wood-it madroño; 5. there and caused-numerator he wood-it there. 6. Pause-word left three, four months, and went he, knowing he going-swollen stomach tree there. 7. Pause-word there and went, knowing he swollen stomach tree there; 8. going stomach tree, and counted he

months. 9. Face completed month, went he there; 10. and-then made-hole he stomach tree there knowing he within demonstrative-pointer man small is man within there. (Ibach 1980, p. 244)

In this Mesoamerican vision, man, flora, fauna, and minerals participate in a comprehensive composite cosmos, allowing for the crossbreeding of humans and trees and the taming of animate stones, while ineffable fates are delivered by the asters (Fig. 9.1).

9.2.2 *The Composite*

We have reviewed certain spatial assemblages of Mexico as *loci* of the virtuous or tragic merging of heterogeneous conceptions of living space, geography, and eventually the world. As a heavy legacy of the laborious Spanish colonization, the composite has become a central theme of the culture- and region-forming processes of the nation. With the following words, Sergio Romano describes the pattern by which the Mediterranean mythology has been transferred into the New World through the miscegenation of diverse traits.

> "[L'espansione dell'Europa nel mondo] è una singolare combinazione di curiosità e arroganza. [...] Nei quartieri stranieri a loro riservati o nelle nuove città che essi costruiscono durante la grande fase della loro espansione, i 'colonizzatori' riproducono lo stile, lo spazio e i principali edifici dei luoghi da cui provengono. [...] Lo stile di questi edifici e i materiali di costruzione sono spesso una dimostrazione della curiosità prensile e mimetica con cui gli europei s'impadroniscono di tutto ciò che li circonda. Ma i palazzi e le case sono collegati l'uno all'altro da una logica spaziale che riflette la cultura politica, civile, estetica dei colonizzatori."[2] (Romano 1998, p. xi)

Although the 'spatial logic' that rules the new cities and towns of the *colonia* is imposed by the colonizer, that logic is implanted into a territorial fabric that is already embedded in the Mesoamerican space. The fact that the existing territorial fabric of the Mesoamerican space is made more of a liquid mythology rather than a solid structure does not deceive us about the depth of its roots. The indigenous spatial logic, in fact, has survived through the centuries of the colonization by virtue of adaptive resistance and metamorphosis. It is surfacing again today in the movements for the recuperation of its cultural values.

---

[2]In English: "[Europe's expansion in the world] is a peculiar combination of curiosity and arrogance. [...] In the foreign districts reserved to them or in the new cities built during the great phase of their expansion, the 'colonizers' reproduce the style, the space, and the main edifices of the places they come from. [...] The style of these edifices and the construction materials are often a demonstration of the prehensile, mimetic curiosity through which the Europeans seize all that surrounds them. But the palaces and houses are related to each other by a spatial logic that reflects the political, civil, aesthetic culture of the colonizers" [translation by the author].

Fig. 9.1 Mixtec nobleman being born from a tree in a pictogram illustrating the myth of the origin from the Codex Vindobonensis or Yuta Tnoho, page 37, dated between the fourteenth and fifteenth century (*photo* courtesy of the Österreichische Nationalbibliothek)

9.2.3 *The Multiple Axes Addressing Man's Presence in Space*

Promising or doomed aspirations to an emancipation of the indigenous are resurrecting from the miscegentated spaces of Nueva España. Various initiatives propose to intellectually and physically reformulate the postcolonial landscape through in situ transcriptions of the processes of domestication, mythological implantation, and remapping of the transculturation processes that have followed.

Among such initiatives, remarkable programs have been started in the open landscapes of the area of Soconusco in the Isthmus of Tehuantepec. Having domesticated the Theobroma Cacao since the early Pre-Columbian era, the communities diffused in the submountain regions of the Isthmus have been living in a pristine equilibrium with the semi-savage ecosystem regulated by the cultivation, processing, and trade of cacao. The Aztecs had colonized a large region around Soconusco to supply their empire north in the Mexico Valley with the finest cacao produce. Under the influence of modernization (which has industrialized the cultivation and relocated the plantations to different continents impoverishing quality) and globalization (which has attracted the younger generations to larger urban centers to fulfill induced necessities), the submountain regions of the Isthmus and the cacao landscape had suffered a diffuse phenomenon of abandonment. Citizens associations and local institutions have recently started to promote a modern agro-alimentary cacao economy of low intensity, locally rooted while globally connected, based on semi-wild growing optimization (in alternative to industrial farming), in situ artisanal produce processing (such as chocolate processing, in alternative to raw cacao trade), and fair product commercialization.

The Ethnobotanical Garden of Oaxaca is another spatial manifestation of similar initiatives, this time within contained precincts. Brought forth by a grassroots initiative started in 1993 under the leadership of anthropologist Alejandro de Ávila and artist Francisco Toledo, the gardens recover the ancient complex of the Santo Domingo convent for civic and cultural purposes. The garden in Oaxaca takes advantage of the extraordinary heritage of both biological and anthropological diversity of its region. The garden's *raison d'être* is poetically narrated by de Ávila in a book that followed the establishment of the garden (De Ávila and Salcedo 2006). The reformulation of a novel space and its bearing mythology in the Oaxacan garden, and in more general terms in the Mexican territory, is plastically molded in processes of domestications organizing man's presence in space along four axes: second nature/productive landscape; symbiotic ecology/man-flora-fauna assemblage; psychic landscape/entheogen-perceived.

The first axis elaborates on the prehistoric domestication of maize and the consequent man/plant mutual interdependence: man feeding on maize grains and maize relying on man's labor to detach its seeds from the cob and reproduce.[3] By stabilizing cochineal on nopal plants, a second domestication is meant to keep the trifold rapport of man, plant, and insect in a munificent equilibrium, yielding the precious dye that has propelled a vast economy.[4] A third domestication links the indigenous Mesoamerican communities to maguey, through a series of taboos governing the extraction of sap and its processing into the traditional fermented drink of *pulque*. The maguey with its derivatives is a central element of the community's diet, economy, medicine, and system of values ruling social dynamics.[5] The fourth axis addresses the subject's perception of the environs, its organization in a conscious *entendement* and the potentials of its psychedelic distortion or expansion.

The ritual use of entheogen plants by native groups in the Oaxaca valley is observed by Richard Evans Schultes since 1939, at the beginning of his research on plant hallucinogens, later completed in the Amazon forest. Schultes's observations, disseminated through his teaching at Harvard, would inspire numerous relevant figures of the psychedelic movements of the 60s. His work is eventually

[3]The domestication of maize by prehistoric populations of the Oaxaca valley-starts from a variety of *teosinte*, a wild grass with a simple fruiting structure composed of loose kernels on a single row common in the region. Through still unclear anthropogenic manipulations, its fruiting structure mutates as radically as to give form to the corn ear we know today. The anomaly of maize stands in the structure of its ear, whose grains are so firmly attached to the cob in multiple rows and wrapped in such a resistant husk of leaves that the plant has lost its ability to autonomously reproduce itself by dispersion of seeds. While man becomes dependent on maize for daily survival, maize becomes dependent on man to detach and sow its grains. The man/maize relationship has been reshaped into a reciprocal interdependence (De Ávila and Salcedo 2006, pp. 16–8).

[4]The nopal-cochineal domestication does not just involve a plant's relationship with man, but is rather performed on a parasitic connection bonding the prickly pear nopal and its plague, the cochineal. A trifold rapport involving man, plant, and insect is reformed into a munificent equilibrium that from cochineal extracts the 'blood of the prickly pear', according to its Nahuatl name, a dye that had colored textiles and artifacts of indigenous peoples. The domestication that stabilizes the parasite over the plant causes the insect to grow bigger and the nopal pads more delicate, dropping their thorns. After the Spanish conquest, the production of 'Spanish red', performed through indigenous labor, is at the basis of the worldwide trade that showers the Spanish crown and merchants with immense wealth (De Ávila and Salcedo 2006, pp. 21–5).

[5]An oral account in Mixtec, recollected in 1970 by linguist Cornelia Mak from a syncretic Christian-animistic community of the Oaxacan highland, attributes to the maguey plant the feelings of a sentient being. The man who cuts the maguey's budding flower stem cannot be the one who scrapes the heart of the plant to draw sap. The man who cuts cannot be the one who has planted the maguey. The man who has planted it can be the one who scrapes. One can set a straw to suck up the sap into a jug. One can cover the scraping with a stone to prevent foxes from sucking the juice. If the scraping changes hand, the maguey stops yielding sap. If a man is hired to cut someone else's maguey, he must burn copal incense and chant to the Virgin of Remedies before others can drink. An array of unfortunate to fatal occurrences expects the ones that do not comply with the rules. The ritualization of maguey-tending activities is aimed at optimizing *pulque* production, but also at harmonizing the social interaction of the social group before and during consumption (Mak 1977, pp. 115–9).

reorganized in 1979 in *Plants of the Gods: Their Sacred, Healing and Hallucinogenic Powers*, co-authored with Albert Hofmann, the creator of synthetic LSD. In this milestone of ethnobotany and ethnopharmacology, the scientific analysis of each plant's chemistry is matched with the anthropologic observation of the social dynamics associated to its use.[6] Once more forming a symbiotic bond, the shamanic use of entheogen plants serves the indigenous communities of the southern Sierra Madre Oriental by favoring the opening of a shared perception of their living space, in return for man's stewardship securing the plants' propagation.

The necessary reassessment of a non-positivistic dimension of reality and reopening of the doors of mystic perception, in combination with a scientific analysis of the world, is a broadly accepted idea today. Achieved through the liberation movements, alternative cultures, and their naïf practices, the idea of a necessary symbiosis with the cosmic forces has permeated the technological structures of our contemporary society, being now conjugated in less ingenuous fashions well beyond any confuse holisms. The Tests for a Symbiotic Matorral project presented in this book[7] aims at cutting a channel through the composite assemblage of the surrounding space, establishing a symbiotic *entendement* of it, and recuperating a glance over the hazardous world, that community with things *autres*, which had once harbored the life of 14 Strength and has then been traded for technique.[8]

9.2.4 Den Wald Zu Bauen

Both consciously and unawares, a collective action has already attacked the formulation of the novel spatial platform for the societal interchange of the post-urban *civitas*. As a consequence, landscape has become the privileged subject of disciplinary fields traditionally distant from each other, such as ecology and aesthetics, as well as, among others, architecture.

The roots of this vast interdisciplinary upheaval go deep into the incrementally mixed character of the space in which we live, of which the miscegenated spaces of Nueva España are a precedent. The digital explosion interweaves the postmodern natural/artificial continuum with a virtual weft (Chu 2004, pp. 74–97) that expands

---

[6]Among other entheogen plants the *salvia divinorum*, or diviner's sage, is one of the rarest psychoactive plants, native to a limited natural habitat in the Mazatec region on the Sierra Madre Oriental. The 'Salvinorin A' contained in the plant is considered as one of the most powerful psychoactive substances in nature. The Mazatecs, a syncretic Christian-animist people, call it María Pastora, as the female shepherd identified with the Virgin. As the plant's natural reproduction through seeds dispersion is almost unknown, its propagation is obtained by means of plant cuttings, which conditions its survival upon human stewardship (Schultes et al. 1992, pp. 164–5).

[7]See Chap. 8 of this book.

[8]See Ernst Jünger's discourse on pervading technique in Chap. 6 of this book.

the sphere of psychic landscape (Lab M4 2015). Recent elaborations on these themes are grafted into the geophilosophical platform, founded onto Heidegger's idea of *'bauen'*, dwelling as 'taking care of' (Heidegger 1993, pp. 343–64),[9] and Jünger's idea of the *'Wald'* (the retreat into the woods).[10] Contemporary geophilosophy has emancipated itself from anarchist individualism through the action of 'molecular collective auto-subjectivization' advocated by Guattari and reformed by the *scuola territorialista* by virtue of the principle of the 'elective belonging' in favor of an operative opening to diversity.

The open task is now that of acknowledging this collective movement, reading its overall trajectory and components, and be ready to contribute to its adjustment towards a higher quality of our physical and intangible space.

9.2.5 Symbiotic Paradigms

The Tests for a Symbiotic Matorral project is also the spatial transcription of a theoretical model aiming at reuniting a disarticulated galaxy of landscape levels and components floating in the international interdisciplinary debate. Through the symbiotic landscape paradigm, this book contributes to the collective challenges of the expanding living platform by proposing a comprehensive understanding of the contemporary space as a composite assemblage, where elements and fragments proceeding from apparently opposite orders coexist.

The compared analysis of theoretical models and applied cases proposed in the research focuses on various levels of the landscape construct: ecological, cognitive, semethic[11] (Farina), metabolic (Baccini), machinic (Mostafavi), territorial (Camporesi), cultural (Sauer), aesthetic (D'Angelo), semiotic (Rowe), social, psychic (Guattari), etc. The proposed critical synthesis reconstructs a model of landscape, reconciling conflictual perspectives presently polarized on 'cognitivist' or 'culturalist' positions: a 'symbiotic landscape paradigm' that describes both ecosystem and place, as congruent components of the same multifocal vision. This assemblage of interdisciplinary components aims at contributing to the reformulation of a possible future for the man/nature relationship where the technical is not independent of the cultural. Systematic research, expression of the technocratic

---

[9]Already expressed by Carlo Cattaneo over a century before Heidegger. See Chap. 3 of this book.

[10]In 'Il filosofo e l'Anarca. Intervista a Ernst Jünger' with Antonio Gnoli, Franco Volpi, Jünger defines the *'Wald'* as "… per me il bosco non è soltanto come per Heidegger il luogo naturale concreto in cui vivono e operano i contadini della Foresta Nera. […] Il bosco è per me soprattutto una metafora: sta a indicare un territorio vergine in cui ritirarsi dalla civiltà ormai segnata dal nichilismo e in cui l'individuo può ancora sottrarsi agli imperativi delle chiese e alle grinfie del Leviatano" (Jünger 2006, p. 54).

[11]For a definition of 'semethic' (semiotic + ethic), language and behavior, word and action, see (Farina 2009).

regime of physical and intellectual production, centered on quantity and automated processing of large amounts of data, has to be mitigated by a subjective research focusing on quality and the overall sense.

9.3 Determining the Attributes of the Novel Urbanity

A fundamental task in the agenda for the Next Landscape is that of determining and evaluating the sets of characteristic attributes of the novel urbanity, pertinent to the diverse conditions and contexts of the contemporary landscape continuum. It primarily means to work on the access of post-urban communities to opportunities and values such as: commerce, information, culture, random societal interaction (Burdett, Sudjic); mobility, congestion, multiplicity (Koolhaas, I. de Solà-Morales); vicarious encounter, implanted mythology, site in land and course of action, habitability (Rowe, Marot, Hunt); control of environment, control of natural systems, interconnectivity, cybernetics (Baccini, Ellis, Lootsma); etc.[12]

9.3.1 Contexts to Start with: Idyllic Global South, Schizophrenic Global North, Welfare Global North, Pseudo-urban Assemblages

Future research work should start from at least four contexts, exhibiting clearly distinct conditions of novel urbanity (or post-urban cityness): (1) the small communities sparsely dwelling in the surviving semi-wild landscapes of Mesoamerica, the 'idyllic global south'; (2) the opposite case of the seminomadic crowds floating in the Asian megaregional mats, the 'schizophrenic global north'; (3) the fully developed and socially civilized metropolitan system of the metropolis of the northern hemisphere, the global north of the welfare; (4) the pseudo-urban population dwelling in the territories of the Mediterranean basin, where the surviving urban space and the preservation of venerable rural and natural fabric coexist with heavy metropolitan infrastructuration, as well as the *terrain vague* of social alienation coexist with European models of safeguarded social space.

The idyllic global south of Soconusco, exhibiting pristine natural systems, semi-wild cacao woods, farm fields, sparse dwellings, and archeological areas, exemplarily represents contexts where, despite the anthropic presence, pristine nature is still dominant and the prevailing geographic platform is that of the *ur-Landschaft.* The landscape of the Isthmus of Tehuantepec is highly determined both culturally and eco-systemically: the ancient regime supported by cacao tree woods is now in decay alongside the ancient relationship of man with nature that used to

[12]See Chaps. 2, 3, and 7 of this book.

support all the fundamental acts of life. All in this context revolves around the possible recuperation of an idyllic relation of man and nature, the perceived proximity of the animistic divine, and the idea of gardening the planet. As Gilles Clément puts it: 'organizing the human territory while attending to living things'.[13]

At the very antipodes, the schizophrenic global north of Tokyo represents the contexts where the geographic platform is constituted by an endless artificial metropolis. The megacity is an artificial monolith grown at the geographic scale: in it, pristine nature has been completely erased and replaced by an artificial nature serving all the material needs of man. There, man wanders over an endless all-providing grid, achieving a technological form of neo-nomadism, similar to the condition evoked by the radical visions of the 70s. Well beyond the metropolitan condition, in the megacity, the fundamental acts are bartered for aesthetic behaviors.[14]

The northern metropolises that we might use as a third reference are Toronto or Montreal, cities supported by a robust economy and efficient infrastructures at the regional scale. But, contrary to analogous cases in the United States, these contexts exhibit a progressive sociopolitical system that contemplates among its priorities that of ensuring its citizens' high standards of welfare and quality of life. The socially civilized northern metropolis is a place where the metabolic exchanges between city and nature are effectively monitored and oriented towards a virtuous equilibrium, tending towards scenarios of cybernetic control over the environment. This context is about endowing a large portion of the community with a convenient living space.

Fourth context, the Mediterranean space is characterized by the coexistence of different orders, among which the permanence of layered territorial fabrics with articulated rural patterns and urban systems. As we have seen earlier, we can retrace the grounding of the contemporary urban and territorial model in the Mediterranean basin back to the middle ages, with the anthropic stratification molding the pristine natural platform. Basically, it is about a hierarchic assemblage regulated by overlapping matrixes and primary components of some monumentality, intended the latter as the capacity to precipitate collective values into space and supply a reference. Recent developments have overwhelmed the historical articulation of the rural realm and suffocated the relatively small urban organisms within hypertrophic proliferations. Only residual historical centers and few recent complexes enjoy the qualities of traditional urban space. Equally limited portions of the countryside maintain those deep-ingrained qualities that endow them with the capacity to produce the long-term renewable wealth that Alberto Magnaghi advocates for.[15] The result is the most complex context of this brief casuistry: a pseudo-urban spatial system, where fragments of rural fabrics and scattered urban nuclei survive, although interrupted, isolated, or functionally neutralized by the overwhelming

[13]See Chap. 6 of this book.

[14]See subchapter 'Of new nomads' in Chap. 2 of this book.

[15]See Chap. 2 of this book.

scale of recent undifferentiated growth, to the point of becoming delusive. As mentioned before, despite its peripheries, disruptive infrastructures, undifferentiated proliferations, a city such as Marseille is representative of contexts still grounded into the history of urban space and its original topography. The *terrain vague* of bleak *cités* looming at a distance coexists with the layered territory at the rim of the Mediterranean, the venerable port city, and the *anciennes hameaux* nestled in the recesses of *Les Calanques*.

Focusing on the determination, comparison, and evaluation of the various attributes of the novel urbanity, the future work spans over a varied geography, much broader than the basic instances described above, and a wide set of interdisciplinary implications, ranging from urbanism to sociology, anthropology, ecology, etc., heading towards the comprehensive notion of life science. This cursory description of an array of contexts and sites is meant to open the discourse about the attributes of novel urbanity.

9.4 Post-urban Agoras, Neo-Monuments, and a Novel Mythology

9.4.1 Urbs, Civitas, Cityness, Landscape: Recharging a Fragmented Territory with a New Sense

We have observed how built systems extend today at the geographic scale of the anthropic continuum. With the built systems, also some of the attributes of the *civitas* (or 'cityness') have expanded. What prevails, however, is the removal of the 'vicarious encounter', the casual interaction between people and spaces offered in the pedestrian occupation of urban public space, source of self-awareness as described by Rowe (1991, p. 59). Burdett equally focuses on the loss of random societal interchange in the new condition.

While the terrains of the post-urban city expand, the urban space disappears along with any trace of *urvus*, the curved trench traced by plough to separate town from rurality and wilderness. The city-*civitas* of the citizens expands, while the city-*urbs*, material platform of *logos* and action, disappears. Arendt[16] says that for the ancient Greeks "the lawmaker was like the builder of the city wall, someone who had to do and finish his work before political activity could begin", to describe the classical coincidence of public space and democracy (Arendt 1998, pp. 192–8). In fact, "before men began to act, a definite space had to be secured and a structure built where all subsequent actions could take place, the space being the public realm of the *polis* and its structure the law; legislator and architect belonged in the same category" (Arendt 1998, pp. 194–5).

[16]See also subchapters 'Athens' on Rossi's reading of Athens and 'In Civibus et in parietibus' on Saint Augustine's *civitas in civibus* formulation, in Chap. 3 of this book.

Today, the architect and the legislator are no longer equated and an unbridged gap has divaricated their fields of action. To avoid plummeting into fragmentation and placelessness, we need to rethink our living space as a continuous landscape. A novel gaze has to recharge today's territorial fragmentation with an overall sense. A novel mythology has to interweave the plot of a landscape carpet of geographic scale: a landscape mosaic composed of forms and processes,[17] both metabolic and semiotic, that is ecological and cultural, formal and functional.

9.4.2 *Life with Neo-Monuments*

The implantation of a novel mythology over the contemporary continuum exacts the reconsideration of the notion of monument and its reinvention in the new context. Our need for novel forms of monument derives, on the one hand, from the definitive demise of its traditional function within the urban and territorial structure and, on the other hand, from the perpetuated necessity for a new civil platform to support the dynamics of democracy. The laws of the formation of traditional urban and territorial structures, anchored to the notion of monument and its physical instantiations, as described by Marcel Pöete, Maurice Halbwachs, and Aldo Rossi,[18] have lost almost any applicability under contemporary circumstances. However, the fundamental monument/residence dynamics, generative of inhabitable topologies, still hold the potential to be reformed into a diffuse democratic platform within today's disintegrated environment.

The contemporary challenge is, then, reinventing the novel *agoras* of the post-urban time: the *agoras* of the expanded *civitas* corresponding to the expanded landscape. Not the flamboyant *agoras* serving 'external forces' (such as the opening and closing hours of the stock exchange inspiring the 'manual for a machinic landscape'[19]) rising in the heart of the privileged metropolis, but rather modest capsules of the quotidian and their interlacing in fabrics and mosaics, networks of empowerment for a diffuse community of non-metropolitanites. The post-urban *agoras* are certainly to be reinvented by enhancing the 'hospitality', that is the spatial quality, of the 'interstices' in-between suburban components, and also the hospitality of suburban infrastructures, such as corridors of mobility and commercial facilities, as Peter Rowe sharply points out.[20]

---

[17]Almo Farina describes the 'visible landscape' as an assemblage where the visible spatial formation of the mosaic is 'coupled', that is coincident, with the observed regime of processes that generate it. In an 'uncoupled landscape', the regime of processes remains invisible. By that, Farina establishes a distinction between the mosaic formations and process metabolisms composing a landscape (Farina 2009, pp. 25–6).

[18]See 'A Critique of the Urban Paradigm and Its Territory' in Chap. 2 of this book.

[19]See 'A Critiques of Landscape Urbanism' in Chap. 2 of this book.

[20]See subchapter 'Interstitial inhospitality' in Chap. 3 of this book.

Fig. 9.2 Suburban vegetable gardens and orchards cultivated by residents in Ikuta, Kanagawa prefecture, in the outskirts of Tokyo's agglomeration (*photo* rp)

However, a new meaning for this idea of hospitality has to be achieved by means of observation and understanding, for it to pervade the expanded dwelling space and reform the canopy of the café at the outskirt of the provincial town, the stairway along the side of a neighborhood library in Torre del Greco, the perfect niche carved in the uppermost corner to provide room for imagination, the modest ecosystem along the ravine dividing a poor neighborhood from a university campus, its surprising floral diversity spanning from *matorral* to *bosque galeria*, the hidden garden of Gibellina, where the dreams of a disappeared generation have been stored, the enclosed orchard where you recognize a redbreast and a bergamot tree, the line of trees that once organized the field pattern and now obsolesces away, as well as the suburban train station of Ikuta in the Kanagawa prefecture, where a traditional shop outside the southern entrance sells tofu in various preparations and the vegetable gardens attended to by the residents conform a hidden rural oasis, a countryside the size of a postage stamp, secluded between retaining walls and the back of the houses (Fig. 9.2).

The challenge is as much about the acknowledgement of the existent quotidian references as it is about the installation of new pieces of what we could call antiheroic 'neo-monuments'. It is that of both recognizing and scattering quanta of collective space in the service of democracy, over the residential expanse of the idiosyncratic continuum, where history and autobiography blur into one another.

9.5 A Final Note

The Mesoamerican myth of 14 Strength narrates of fates ineffably administered by asters, whereby the sun rises and for no reasons kills the man born from a tree. The opening quotation of this book, from Giacomo Leopardi's *Dialogue between Nature and an Icelander*, offers a western figuration of the human condition before nature that does not result too dissimilar from that of the Mesoamerican myth: "If I by chance exterminated your species, I should not know it"—says Nature.

Man is proliferating on the planet as an infesting species and has long adopted a parasitic attitude. If control over the number of the human population is not enacted and the parasitic behavior is not turned into symbiotic, a new aster will probably rise, from nature or from technique, to hit the species.

References

Arendt H (1998) The human condition. The University Press of Chicago, Chicago [1958]

Chu K (2004) Metaphysics of genetic architecture and computation. Perspect Yale Architectural J Building Codes, n.35

De Ávila A, Salcedo C (2006) The thorn and the fruit. Plants from the ethnobotanical garden of Oaxaca. Oaxaca, Artes de México

Farina A (2009) Ecology, cognition and landscape. Springer, New York [2006]

Heidegger M (1993) Building, dwelling, thinking. In: Basic writings. Harper Collins, San Francisco [1971]

Ibach Thomas (1980) The man born of a tree: A mixtec origin myth. Tlalocan (UNAM) 8:243–247

Jünger E (2006) Il filosofo e l'Anarca. Intervista a Ernst Jünger. In: Volpi F, Gnoli A (eds) L'ultimo sciamano. Bompiani, Milano

Lab M4 (2015) Mythomanias: a series of psycho-architecture case studies. In: Lacadée C, Roche F (eds) Punctum books

Mak C (1977) Maguey tapping in the higland mixteco (sic.). Tlalocan (UNAM) VII: 115–9

Romano S (1998) Preface. In: Benevolo L, Romano S (eds) La città europea fuori d'Europa. Libri Scheiwiller, Milano

Rowe P (1991) Making a middle landscape. The MIT Press, Cambridge

Schultes RE, Hofmann A, Rätsch C (1992) Plants of the Gods: their sacred. Healing Arts Press, Healing and Hallucinogenic Powers

Sudjic D (2007) Theory, Policy and Practice. In: Sudjic D, Burdett R (eds) The endless city. The urban age project by the London school of economics and the Deutsche bank's alfred herrhausen society. Phaidon Press, London

Afterword

A Borgesian Tabula: The Symbiotic/Composite as an Epistemological Rupture and a New Paradigm in Landscape Research

This book by Roberto Pasini marks a point of methodological heterogenesis in the context of the consolidated landscape literature: a conceptual alteration that substantially changes our relationship to the project, primarily intended as a cognitive rather than a spatial practice.

After dismantling the traditional landscape theory, Pasini recomposes a non-dichotomous edifice of *quanta* of deterritorialization, a non-anthropocentric 'cognitive machine' that allows for a continuous becoming of the contents. By weakening the strong figures and abandoning Manicheism in favor of a conceptual nomadism, the ecological thinking appears in a most complex and experimental conjugation. The work combines a geo-philosophical and relational apparatus with the purely scientific perspective, defining a new notion of 'composite symbiosis' that finds its transdisciplinary roots in the Deleuze-Guattarian 'becoming-' and Almo Farina's 'cognitive landscape'. In an absolutely original way, Pasini elaborates the combination of those opposed positions into a theoretical definition first and then transfers it into the design world, breaking rules and taxonomies with more wandering contents. As a 'body without organs', in fact, the new paradigm performs a reorientation of the precepts of architecture towards other taxonomies of deterritorialization and reterritorialization that generate a new 'play of planes': the 'plane of consistency' of the becoming, on which one commeasures the heterogeneous and Aeon's and Kronos' time (Deleuze 1969) overlap. It is a new theoretical space that entails an epistemological rupture in the world of landscape design, embodied in the *matorral* installation in the Sierra Madre Oriental.

R. Pasini, *Landscape Paradigms and Post-urban Spaces*, The Urban Book Series,
https://doi.org/10.1007/978-3-319-77887-7

The disciplinary value of the text stands in fact precisely in the possibility of defining and experimenting a new untrodden ramification of landscape ecology capable to resolve, represent, and project the global and particular space—at once heterogeneous and composite—in a symbiotic fashion, that is non-definitive, relational, 'miscegenated' and non-anthropocentric. The value of the union of the heterogeneous in an ecological perspective is thus the distinctive trait of this book, in which the heuristic and spatial principles of the project are redefined alongside the tools and glossary of landscape. The *matorral* project simultaneously is, and this is again original, a theoretical hologram and an integral immersive space. In it, biodiversity, sensoriality (implying the drop of vision as a synthetic and unifying act) and both semantic 'verticalization' and geographic 'surfaceness' imply the genesis of a theory not simply ecosophic, but rather based on a relational osteology, on the paradox of the open and dynamic structure, that is a 'body without organs'.

About the Symbiotic/Relational

In constructing his theoretical basis, Pasini tends to the overcoming of the exclusivity of the scientific, allowing for a diagonalization: the mechanical and neo-deductive model of pure ecology is overtaken as well as the purely narrative model and the conventional disciplinary axioms are abandoned.

By passing from dialectic to dialogic, the text advocates for the elimination of binary thinking in favor of real conditions and opts for a third way: Morin had defined the "seventh way of complexity" as "the way of the crisis of closed and clear concepts, that is the crisis of clarity and separation within explanation" (Morin 2008). Manichaean dichotomies as architecture/city, city/countryside, etc., or conceptual oppositions such as order/disorder, full/empty, can be overcome for landscape science to work with differential, simultaneous, and non-oppositional contents. The play switches then from dialectic to a dialogic rapport among agencies: from a thesis/antithesis order to an open system without final solutions and discontinuities.

By eliminating the designative, the monadic and the self-referential, this theoretical project proposes a critique of the division of matter in separated bodies and systems that results either in the designed object's hyperthelia, epitomized by Koolhaas' production, or in the exclusive coherence of homogeneous elements. The reflection leads to the idea that the relationships between the project and the world exclude the project's and its process' claims to a substantial and conforming value. The ancient principle of interrelation among things is rather reinstated. In *La Monadologie*, Leibnitz had asserted that "tout corps se ressent de tout ce qui se fait dans l'univers" and that "tout présent état d'une substance simple est naturellement une suite de son état précédent, tellement, que le présent y est gros de l'avenir. Nous n'avons que des définitions nominales provisionnelles" (Leibnitz 1714).

This spatially relational and timewise provisional mandate of the project conducts to a multiplication of the characteristics of the space, which loses any principle of homologation and disconnects itself from the systemic theory, rather embracing the metaphor of the rhizome (Deleuze and Guattari 1980). And the elements of the symbiotic project, in its interpretation of the 'becoming-', present themselves in the following acceptations: a reflection on the value of the horizontal surface as opposed to the "metaphysical [conception] of the ground" (Deleuze and Guattari 1980, p. 18); the possibility to work in a *topos* of this relational system without affecting its processual as opposed to substantial value; the substitution of identity with the 'singularization of multiplicity' (Deleuze and Guattari 1980, p. 555); considering the process of the project as 'a' moment in-between the conditions and not as 'the' moment and obliterating the concept of beginning and end; the idea of permanent deterritorialization of the contents and concepts. In this sense, the design is treated as a method of knowledge, as a "network of connections among the facts… the things of the world" (Bergson 1896), conscious of the fact that "knowing is inserting something into the real; it is then transforming the real" (Calvino 1988).

Pasini submerges the project into the flow of things, into a phenomenology that includes the real and the imaginary, and strips the things of their parodic and imitative value, passively received in common use and direct phenomenology, exacting from the reflection on the landscape new duties and an alteration.

The Taxonomy of the Relational: The Symbiotic/Composite

If, from a theoretical point of view, the text innovates the apparatuses of landscape as a cognitive practice, the terms 'symbiotic' and 'composite' articulate the methodological consequences of that cognitive practice. 'Symbiotic' and 'composite', as reconnaissance tools, seem to evoke Glissant's *éclat* by summoning an operative project that fluctuates between theory and practice (Glissant 1990). In line with Farina's notion of 'semethic' (meaning, sign, ethic) (Farina 2011), the two terms retain a polysemous character and invite to explore routes of rupture. The two terms, at once conceptual and operative, are then the interface between 'things' and 'thought', exactly as Carroll's "eating and talking" (via Deleuze) (Deleuze 1969), and synthesize both the drop of the dialectical contents and the opportunity to institute a 'temporarily foundational' language. The symbiotic and the composite therefore become the pivots of a new epistemology of the project, *mot de passe* to a modality that allows the author to advance experimentally. The language becomes a projective, experimental, multiple and concrete instrument, whose unexplored design 'figurations' build the place of interchange between thought and phenomenon, between deterritorialization and reterritorialization: an ever-reformable interchange *vis-à-vis* diverse contexts, defined from the beginning as open and potential.

Overcoming the positivist/idealist definitions of ‘symbiosis’ and the eclectic/organicist definitions of ‘composite’, the proposed model only retains the sole meaning of the coexistence of organisms of diverse species. Beyond the subsistence of the organisms involved, the discourse is about the expansion of the material and immaterial faculties of living beings through reciprocal contamination and modification. And, within this transfer, Pasini abandons any Linnaean temptation of pure ecology, carrying the project onto a board that allows for the coexistence of micro and macro, man and animal, natural and artificial. Even the hybridization of different conditions of homogeneous categories, such as natural/natural and cultural/anthropological, finds in the dialoguing plane of this theory and design model a new relational space complex *per-se*: a rupture that overthrows the plane of the discourse about the project.

Eventually, the symbiotic is a transversal and potentially universal method: beyond the fusion of oppositional polarities, it is the reconciling principle of distinct natures and elements in the local as well as universal project, intended as the conceptual plane recollecting the different and, by that, transforming the thought about the project. It is in this relational movement across the heterogeneous and also the homogeneous that the method finds its significance through the determinations of ‘miscegenation’ and ‘garden’ as metaphors of the project.

The transfer of the composite/heterogeneous into the notion of miscegenation recalls again Glissant’s *Poetics of Relation*. By virtue of miscegenation, the axiology of the proposed system becomes not only theoretical/methodological but also anthropological/ethical, instituting a new space of signification for man’s diverse cultures. By representing through miscegenation a vision of the world as “minoritarian becoming” (Deleuze et al. 1985), the proposed model introduces a new kind of relation to the other—poetically evoked through the figure of Josephine Baker—and overpasses the demon of purity of systemic theories to open the opportunity for a relation between “the same and the different” (Deleuze 1968). The purported principle of Western authenticity no longer appears as an absolute image, rather replaced by a terrain entailing alterity and difference: a “unity that goes by the diversity” (Deleuze and Guattari 1980, p. 17). It is through this idea of miscegenation that the entire second session of the book acquires a substantial meaning: the possibility of a different relation among cultures and a new culture of the project, in which “the ethos of the nomad and pilgrim becomes the ethos *par-excellence*” (Lévy 1994) or the final embodiment of the “chaos/world” occurs.

The Matorral Project as a Theoretic Hologram: The Becoming-Demon

The notion of miscegenation becomes spatial in the *Tests for a Symbiotic Matorral* project, intended as both the place of the Deleuzian ‘becoming’ (a geographical phenomenology) and a specific ecotope within a ‘cognitive landscape’.

By materializing a rhizomatic structure, the *matorral* project transfigures the dynamics of the local ecological resources into intersubjective parameters mediating between man and nature. The central section, called the *Glade of Becoming*, is crossed by four transversal axes that introduce to the condition of various animals by letting the visitor experience through their sensorial faculties. The longitudinal path of the 'becoming myth/becoming demon' lines up the transversal axes in an animal multiplicity. The distribution of resources/signals over the *Glade of Becoming*, in which ecosystemic traits of flora and fauna coincide with the cultural values of man, deploys a multifocal and anthropological vision culminating in a total experience of 'the other', up to the point that, across the trail, the human metaphorically transmutes into a demon as symbiotic and atavistic as werewolves and shamans (see Chap. 8 in this book). The installations on the glade, in fact, provide a tangible experience of the perceptive and cultural deterritorialization advocated for at the beginning of the book, pointing to a new 'anthropological organization' alternative to current social structures (family, community, etc.).

The project's syntax leads from the experience of the single (the subject/visitor) to that of the many (nature/animals). As places designed at variable intensities, the installations unite the "sensorial perceptions of the observer with the functions and the fundamental acts of the various ecological species" (Guattari 2000, p. 34): in three sections—perceptive, experimental, informative—respectively called fauna-trails path, glade of becoming and flora path, ushered by a portal/threshold and concluded by the contemplation circle, man and animals intersect their longitudinal and transversal trajectories experiencing each other. On the glade, the planes of transcendence and immanence overlap on each other over the general plane of consistency.

Regulated by the Deleuze-Guattarian anti-taxonomy of the animal world—domestic, mythic, demonic—the visitor experiences this very passage moving from one post to another in a metamorphic progression attaining to the physical and the imaginary. The subject proceeds deeper and deeper through multiple thresholds on the path of his transformation to finally reach the play of shadows and transparencies of the central totem, a place of reconciliation of the multiple that reconnects the dichotomies of man/animal, archetype/totem, series/structure, imagination/comprehension, resemblance/relation. By dropping anthropocentrism, the visitor crosses a sequence of displacing spaces where, dwindling and expanding, rising and squatting, watching and smelling, like in Alice's world, the names are lost and the cogitating ego becomes the other. This process is formalized through open devices, which innovate the compositional regime of the tradition to let in Guattarian impermanence, weakness, fluidity, abdication to aprioristic figures of order in favor of a mobile, multisensorial composition of space. In fact, the composite symbiosis occurs by virtue of those devices: as the outcome of a heuristic principle, the project is the topological manipulation of a conceptual and experiential process that produces an anthropological transformation by way of space.

Despite its formal unrepeatability, this theoretical hologram becomes a general manifesto for a post-urban space in which landscape theory is deployed as a way of global cognition: in this space, symbiotic and composite, real and imaginary, natural and human, constantly reformulating itself, the project finds its own mutable sense in relation with the world. On what could be described as a Borgesian *"tabula"*[1], Pasini seeks a new modality for landscape theory and design and for the project as a category *in-se*: a new, efficient epistemological model that allows for a project universal and particular, global and specific, far removed from common approaches, systemic or ideological, that plague the science of landscape.

Milan, 2018 Elisa C. Cattaneo, Ph.D.[2]

References

Bergson H (1896) Matière et mémoire - Essai sur la relation du corps à l'esprit. Félix Alcan, Paris
Calvino I (1988) Six memos for the next millennium per il prossimo millennio. Harvard University Press, Cambridge
Deleuze G (1968) Différence et Répétition. PUF, Paris
Deleuze G (1969) Logique du Sense. Gallimard, Paris
Deleuze G, Guattari F (1980) Mille Plateaux. Éditions de Minuit, Paris
Deleuze G, Guattari F, Maclean M (1985) Kafka: Toward a minor literature: the components of expression. New Literary Hist 16(3):591–608
Farina A (2011) A biosemiotic perspective of the resource criterion: toward a general theory of resources. Biosemiotics (1):17–32 (Springer)
Foucault M (2005) The order of things (trans: Tavistock/Routledge). Routlege, London, New York [1966]
Glissant É (1990) Poétique de la Relation—Poétique III. Gallimard, Paris
Guattari F (2000) The three ecologies. The Athlone Press, London [1989]
Leibnitz GW (1898) The monadology and other philosophical writings (trans) Latta R. Oxford University Press, Oxford [1714]
Lévy P (1994) L'Intelligence collective. Pour une anthropologie du cyberespace. La Découverte, Paris
Morin E (2008) La Methode, vol. 1. Seuil, Paris [1977–2006]

[1]In the Preface to *Les mots et les choses: Une archéologie des sciences humaines*, Michel Foucault states that his book first arose out of a passage in Borges enumerating the surprising taxonomy of animals from 'a certain Chinese encyclopedia'. The Chinese taxonomy breaks up "all the planes with which we are accustomed to tame the wild profusion of existing things [...] What is impossible is not the propinquity of the things listed, but the very site on which their propinquity would be possible. [...] What has been removed, in short, is the famous 'operating table': [...] the nickelplated, rubbery table swathed in white, glittering beneath a glass sun devouring all shadow—the table where, for an instant, perhaps forever, the umbrella encounters the sewing-machine; and also the table, a *tabula*, that enables thought to operate upon the entities of our world, to put them in order, to divide them into classes, to group them according to names that designate their similarities and their differences—the table upon which, since the beginning of time, language has intersected space." (Foucault 2005, pp. xvi–xix)

[2]Adjunct Professor, DAStU, Politecnico di Milano.